AF328019

TRENDS AND PERSPECTIVES IN PARASITOLOGY: 1

QL
757
.T73
v.1
1981

TRENDS AND PERSPECTIVES IN PARASITOLOGY

I

EDITED BY

D. W. T. CROMPTON AND B. A. NEWTON

CAMBRIDGE UNIVERSITY PRESS

CAMBRIDGE

LONDON   NEW YORK   NEW ROCHELLE

MELBOURNE   SYDNEY

Published by the Press Syndicate of the University of Cambridge
The Pitt Building, Trumpington Street, Cambridge CB2 1RP
32 East 57th Street, New York, NY 10022, USA
296 Beaconsfield Parade, Middle Park, Melbourne 3206, Australia

© Cambridge University Press 1981

First published 1981

Printed in Great Britain at the
University Press, Cambridge

British Library Cataloguing in Publication Data

Trends and perspectives in parasitology. – 1
1. Parasitology – Periodicals
574.5′249′05   QL757   80-42159
ISBN 0 521 23821 8 hard covers
ISBN 0 521 28242 X paperback
ISSN 0260-6763

CONTENTS

PREFACE

Parasitology seems to have become a rather restricted subject with most stress being placed on the study of host–parasite relationships involving protozoa, helminths and certain arthropods. Even within their own specialization, workers acknowledge the problem that they have in trying to keep up with the literature and many feel that it is impossible to read more widely and learn about discoveries in related disciplines. We believe that the series of short articles, now known as TRAPS (*Trends and Perspectives in Parasitology*), which are being published in *Parasitology*, will help not only to transmit new ideas and points of view to specialists working in the same field but will also lead to the sharing of results and opinions from more distant disciplines. High standards of teaching are more likely to be maintained if those who participate are offered news of recent developments in branches of science in addition to their own.

Contributors to TRAPS are given a difficult brief and the editors hope that readers will bear this in mind if gaps in the coverage of a topic are obvious. TRAPS are not intended to be comprehensive reviews; authors are asked to present a personal view about their own interest in a subject. The articles are intended to (1) focus attention on developing and growing points in parasitology, (2) identify new and old problems in need of research, (3) suggest how results from one area of biology might have some bearing on current studies in parasitology and (4) present teachers and students with concise summaries of recent discoveries and new theories.

We very much hope that this series of short books containing collected reprints of articles from *Parasitology* will prove to be useful and stimulating to all who read them and we thank the contributors for their efforts.

Cambridge, December 1980
D. W. T. Crompton
B. A. Newton

CONTRIBUTORS

C. D. H. BOARER is a Senior Scientific Officer at the Central Veterinary Laboratory, Weybridge. His work has involved endocrinological studies of sheep and cattle and investigations of contagious bovine pleuropneumonia and East Coast Fever. These diseases were studied during spells of work in Kenya at EAVRO and recently ILRAD.

A. D. IRVIN qualified as a veterinary surgeon in 1963 and, after 3 years in general practice, joined the Department of Parasitology at the ARC Institute for Research on Animal Diseases. He worked for several years for EAVRO at Muguga in Kenya and is currently based at ILRAD where he is studying theileriasis with particular emphasis on the potential use of *Theileria parva* sporozoites for vaccination of cattle against East Coast Fever.

ANN M. LACKIE graduated at the University of London and obtained her Ph.D. degree from Cambridge where she stayed for some years as a Research Fellow of New Hall and Assistant in Research in the Department of Parasitology. Like her husband, she is a Lecturer at the University of Glasgow where she has made considerable progress in extending our knowledge of invertebrate immune responses.

J. F. RYLEY claims that his scientific career has been shaped by following his interests and being fearful of getting in ruts. He graduated in biochemistry and then worked for his Ph.D. degree on trypanosomes which have held his attention for 30 years. He started his professional career with ICI where he has worked ever since on chemotherapeutic and biochemical aspects of trypanosomes, malarial parasites, trichomonads, amoebae, piroplasms, leishmanias and coccidia. He is now finding a new lease of life working on antifungal chemotherapy.

LANI S. STEPHENSON is a visiting assistant professor in the Division of Nutritional Sciences, Cornell University, New York. Her research work is concerned with human nutrition and she has made significant contributions to knowledge of lactose tolerance. Since 1975, she has been investigating the impact of parasitic infections on the health and nutrition of rural people in Kenya where she has spent much time in attempting to relieve young children from the stress of roundworm infection.

H. T. TRIBE is a soil microbiologist who became interested in nematodes through the work of the plant nematology group directed by F. G. W. Jones in the old Cambridge School of Agriculture. He has worked almost without interruption in Cambridge since 1951 and is currently investigating the basis of the ecology of microfungi in the soil.

DAVID WHARTON After graduating and completing his doctoral studies at the University of Bristol, he held an SRC Post-doctoral Fellowship for two years at University College, Cardiff. His research interests include the structure and function of nematode egg-shells and the survival of freezing and desiccation by the infective larvae of trichostrongyles. His work on larval survival is being undertaken at the University College of Wales, Aberystwyth.

Parasitology (1980), **80**, 189–209

1

Recent developments in coccidian biology; where do we go from here?

JOHN F. RYLEY

Imperial Chemical Industries Ltd, Pharmaceuticals Division, Mereside, Alderley Park, Macclesfield, Cheshire SK10 4TG

(*Accepted* 5 *September* 1979)

INTRODUCTION

In the 20 years that I have been involved in coccidiosis research there have been major advances in our understanding of the parasite at both the macroscopic and the microscopic level. Histochemical and biochemical studies have taught us much, particularly concerning the seemingly indestructible oocyst. Immunological investigations have highlighted the amazing specificity of the immune response and the complexity of this whole field. Notable advances have been made in the practical control of coccidiosis by chemotherapy. Ultrastructural studies have revealed a whole new world of fascinating complexity. Indeed ultrastructural studies of a variety of organisms of obscure phylogeny gave the first clue which ultimately led to the inclusion of *Toxoplasma, Besnoitia, Sarcocystis* and others in the coccidial fold – albeit with a different type of life-cycle, and species and tissue specificity than say the *Eimeria*. Yet over the last few years, I feel that there has been some stagnation of ideas and some loss of purpose in the research effort in the coccidiosis field. It is the purpose of this review to highlight just five areas of personal interest, and to suggest lines of research which might prove interesting – and profitable. My attention will be focused mainly on chicken coccidia, but in a few instances will stray to mammalian coccidia.

HOST-PARASITE RELATIONSHIPS

A parasite requires a host before it can live and multiply, and in the case of the coccidia there is a remarkable degree of specificity both in regard to that host and to the part of that host in which development can take place. It would be interesting to know more about the basis of this specificity – to know to what extent the host behaves passively towards the parasite, providing suitable conditions, nutrition and so on just because it is a suitable host, and responding to the damaging effects of the parasite in the way it would whatever the cause of the damage, or to what extent the host obediently responds to the parasite at the specific direction of that parasite. We can consider the relationship between host and parasite at the macroscopic, whole body level or at the microscopic, cellular level; we can consider the relationship from the anatomical and structural point of view or alternatively from a biochemical or physiological aspect. Some of the host–parasite relationships of the coccidia have been considered by Crompton & Nesheim (1976) in a review covering a variety of intestinal parasites, while I have discussed some of these relationships in a review directed at the harmful effects the coccidia have on the host (Ryley, 1975). Two groups are currently doing particularly interesting work in this area, and it is their work which I now wish to consider.

Ecology of the coccidia

Eimeria tenella, among the chicken coccidia, is unusual in that it normally only develops in the blind gut or caeca rather than in the mainstream of the small and large intestine. Clarke & Crompton (1978) have been investigating the structure and function of the caeca in relation to the suitability of this environment for *E. tenella*. It is surprising what gaps there are in our basic knowledge of the anatomy and physiology of the chicken gut; Clarke's first paper is thus devoted to the structure of the ileo–caeco–colic junction (Clarke, 1978*b*). The caeca seem to be periodic in their filling and emptying activities. Thus, normal caecal faeces are only produced at certain times of the day during a 12 h light/12 h dark cycle, and never at night, and caecal filling seems to be by retrograde peristalsis from the large intestine, presumably at times related to those of evacuation. Although it is interesting to speculate on ways in which different coccidia determine their preference for different parts of the small or large intestine – as I have done in the next section – it is even more difficult to envisage how sporozoites of *E. tenella* get into the caeca. Do they do so only if they happen to be passing the caecal openings when material is being forced in, do they linger in the large intestine until such time as retrograde peristalsis occurs – and if so, how do they manage to linger there rather than get passed out in the faeces directly – or do they have a mechanism for entering the caecum irrespective of whether the organ is emptying or filling or not? Given adequate chemotactic or other directional information, it is not impossible to imagine them making their way into the caeca by their own efforts. If, on the other hand, entry is passive along with other intestinal contents, how is it that sporozoites of other species which have excysted, but failed to invade the upper reaches of the intestine, do not likewise enter the caeca and develop? Clarke & Crompton (1978) observed that the distribution of development of *E. tenella* within the caecum was not uniform. Second-generation schizogony and the attendant haemorrhage was found mainly in the middle part of the caecum, while gametogony was in the middle to distal regions; neither stage developed appreciably in the neck regions. Is this distribution the result of a particular preference on the part of the parasite and expressed by a deliberate migration, or is it merely the result of flow patterns of the caecal contents and a reflection of physical conditions? It is not in the interests of a parasite to be particularly pathogenic; if *E. tenella* exerts its pathogenicity and kills its host at the close of second-generation schizogony, it is unable to complete its life-cycle to produce the oocyst and ensure its survival. There has been a fair degree of interest in the effects of *E. tenella* and other coccidia in gnotobiotic chicks and in chicks with one or several species of associated bacteria (see review by Ryley, 1975). The presence of certain bacteria results in more severe pathogenic symptoms, which in some cases may be due to stimulation by the bacteria of the development of greater numbers of coccidia. There is obviously much scope for further study of interactions between the coccidia and other organisms in the same habitat.

The caeca are blind parts of the intestine with no continuous through-flow of gut contents. Although attached to the lower reaches of the intestine at the junction of the ileum and colon, they lie in the body cavity in close proximity to the duodenum, and indeed share the blood supply of some of the upper parts of the intestine. Much work on site specificity could be done by someone with surgical expertise; because of the gross relationships of the caeca to the other parts of the intestine and their blood supplies, it should be possible to insert segments of caecum into the duodenum and upper small intestine, and likewise segments of upper intestine into the caeca to see whether specificity is controlled by the actual tissues or rather by their location. Ikeda (1957) pioneered this type of approach (as well as factors influencing caecal evacuation), but his surgery was limited to attaching caecal pouches to the duodenum and upper small intestine – a situation which would incorporate no mechanism for emptying and filling – rather than inserting patent segments into the intestine.

Although caecal evacuation in normal birds occurs according to a discrete rhythm, with no caecal faeces at all during the night (although intestinal faeces are voided day and night), during the haemorrhagic phase of *E. tenella* infections, this rhythm is completely upset, with caecal evacuation taking place at frequent intervals (Clarke, 1978*a*). This disturbed rhythm is followed by a period of caecal inactivity for up to 5 days before the normal rhythm is re-established. It is of considerable interest that oocysts, produced in the caeca, are excreted – in intestinal faeces if necessary – whether or not caecal evacuation is taking place.

Host–parasite relationships

Intestinal cells harbouring coccidia undergo traumatic changes as the parasite grows and matures, being finally destroyed when schizont rupture takes place. Some of the effects must be due to the purely physical forces exerted by a rapidly and considerably growing parasite, but some of the changes are possibly mediated in specific obedience to the growing parasite. Fernando and her associates in Guelph have been seeking to quantify some of the changes produced in the host cell. When chick intestinal cells are infected with second-generation schizonts of *E. necatrix*, the host cells undergo extensive hypertrophy, with an eventual 5-fold increase in surface area; there is at least a 4-fold increase in the DNA and RNA content of the host cell nuclei; the plasma membrane of the host cell becomes refractory to fracture by mechanical shearing, hypotonic shock or ultrasound; and eventually extensive autolysis of the host cell cytoplasm takes place. It is interesting to note that the rate of DNA augmentation after infection is neither synchronous nor continuous in all of the infected cells and actual values in individual cells vary considerably (Fernando, Pasternak, Barrell & Stockdale, 1974). In addition to increased host cell DNA synthesis in the case of *E. necatrix*, Fernando and her colleagues observed an increase in the number of nucleoli from 1 or 2 to 2 to 5 or 6; nucleoli also became more distended. Chemical studies on isolated cell nuclei (Fernando & Pasternak, 1977*b*) confirmed the increased synthesis of both DNA and RNA by the host cell during parasite development.

It is not clear whether cells with a proliferative capacity continue to synthesize DNA but are prevented from dividing because of the presence of the parasite, or whether the excessive DNA synthesis is in response to direction from the parasite to synthesize nuclear material for its benefit. It could well be in the interest of the parasite to inhibit host cell division to allow completion of its own development. Any direction from the parasite must be very local, since adjacent uninfected cells show no signs of nuclear hypertrophy. That the effect may well indicate the inability of the cell to divide in spite of the proliferation of nuclear material is suggested by observations on the sexual stages of *E. maxima*. These parasites, by contrast, develop in epithelial cells of the villi, i.e. in cells incapable of reproduction. In spite of the fact that more parasite DNA is synthesized in the microgamonts of *E. maxima* than in the second-generation schizonts of *E. necatrix*, no nuclear hypertrophy was found by Fernando *et al.* (1974) in the host villus cells.

Intestinal cells harbouring second-generation schizonts of *E. necatrix* become resistant to mechanical shearing, hypotonic shock and ultrasound; the membrane is, however, susceptible to such forces after trypsinization. These differences from non-infected cells allowed Fernando & Pasternak (1977*a, b*) to devise methods for isolating infected cells from homogenates of intestinal scrapings and subsequently isolating host cell nuclei. In a recent paper, Thompson, Fernando & Pasternak (1979) describe the isolation of plasma membranes from infected cells and also purified membranes from isolated parasites. The lipid bilayer matrix of the membranes of normal cells is in a liquid crystalline (fluid) state. Wide-angle X-ray diffraction techniques have been used to show that during the later stages of parasite maturation, the host cell plasma membranes acquire increasing proportions of gel-phase lipid. Parasite membranes show a normal liquid-crystalline state. During the later stages of infection, the cells became stainable with Trypan blue, indicating a breakdown in the permeability of the host cell plasmalemma.

It is to be hoped that these studies on host cell changes consequent on parasitization will be continued, and that it will be possible to determine what is indicative of degeneration and destruction and what if anything represents parasite-directed modifications carried out by the host cell for the benefit of the parasite.

FROM OOCYST TO TROPHOZOITE

To me, the most fascinating part of the coccidian life-cycle is not the multiplicative phase that takes place in the tissues of the host, but rather what happens to the parasite once it leaves its definitive host in the form of an unsporulated oocyst until the time a new cycle of multiplication is initiated by sporozoite invasion of epithelial cells. Just how does the oocyst keep alive during a prolonged period of dormancy? What is it about the oocyst wall which makes it resistant to virtually any disinfectant you like to name – and yet allows it to become permeable to the appropriate stimuli for excystation? What causes the sporozoite to

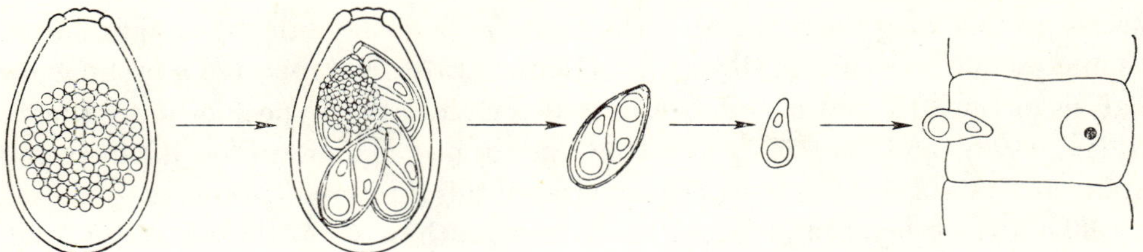

leave its sporocyst and oocyst with such a display of frantic activity? After all it
has no appendages or other visible means of locomotion. How does it find its way
through a mass of digesting intestinal contents – without itself being digested –
to just the right location in which it can develop? And having found the right cell,
without anything to push against or with, how does it manage to invade that cell
and thus initiate the parasitic mode of life?

Oocyst wall

Contamination of premises with coccidia is very difficult to control. Virtually no
common disinfectant – such as is used for the control of bacterial, viral or fungal
organisms – has any effect on the coccidial oocyst. Indeed in the laboratory it is
standard practice to store oocysts in solutions of potassium dichromate or dilute
sulphuric acid, which prevent the growth of extraneous micro-organisms but
maintain the coccidia in robust health. Only small molecules such as methyl
bromide, ammonia and carbon disulphide are able to penetrate the oocyst wall
and kill the contents. In the poultry house situation the first two are difficult to
apply and keep in contact with the oocyst for an adequate period of time, apart
from the fact that they are dangerous to the operator. Carbon disulphide, in
addition to being a toxic hazard to the operator, is also highly inflammable; more
than one poultry establishment has been rid of coccidia following the use of carbon
disulphide by virtue of the fire which ensued!

We made some preliminary observations on the chemical nature of the oocyst
wall (Ryley, 1973). The outer layer, consistently representing about 20 % of the
wall on a dry weight basis, was removed by dissolving in 10 % (w/v) sodium
hypochlorite solution. Due to the large amount of inorganic matter, chemical
analysis was difficult. It did, however, appear to contain carbohydrate, and a
protein characterized by a high proline content and the absence of basic amino
acids. Nyberg & Knapp (1970*b*) had already shown with the electron microscope
that the outer layer of the oocyst wall could be removed by exposure to 2·6 %
sodium hypochlorite solution for 15 min, and Vetterling, Madden & Dittemore
(1971) had confirmed this in a beautiful study with the scanning electron microscope. Stotish, Wang & Meyenhofer (1978) have studied the chemical nature of
the walls of oocysts purified by a treatment which involves exposure to 5 %
sodium hypochlorite. They claim, by contrast, that hypochlorite treatment in no
way affects the intact oocyst wall, but the electron micrographs they present do
not show enough detail to substantiate this claim. *E. tenella* does not have a
micropyle in the normal sense of the term; *E. magna* from the rabbit has a very
prominent one, and the effect of hypochlorite can be seen more readily in this

species, as following the removal of the outer layer of the wall with hypochlorite, the micropyle also vanishes (Ryley & Robinson, 1976). We reported a preliminary analysis of the 80 % of the oocyst wall that remains after hypochlorite treatment (Ryley, 1973), and Stotish *et al.* (1978) performed a much more detailed and careful analysis of this fraction. They envisaged this (inner wall) as being composed of a 90 nm thick layer of glycoprotein covered with a 10 nm thick layer of lipid. The lipid layer appeared to contain large amounts of fatty alcohols, particularly hexacosanol, and appreciable amounts of phospholipid. The outer hypochlorite-soluble layer of the oocyst wall may have some structural significance, but it is this inner, lipid-covered layer which protects the oocyst from chemical attack. From the scientific point of view it is of interest to study this wall further to determine just how injurious substances are excluded, and yet whatever gives the stimulus for excystation can penetrate. From the practical point of view, a method of dealing with this chemical barrier to make the oocysts more susceptible to normal disinfectants could be a real breakthrough. Stotish *et al.* (1978) say 'that the outer lipid layer can be removed in certain non-polar solvents, resulting in increased permeability of the oocysts to aqueous reagents' – but they give no details.

Excystation

Assuming environmental conditions are favourable, the oocysts shed by the host undergo the process of sporulation. Over a couple of days or so nuclear division is followed by organelle production and the formation of sporozoites enclosed in sporocysts. The process is strictly aerobic, and is supported by catabolism of both carbohydrate and lipid. Once sporulation is complete, however, oocyst respiration and metabolism fall to a barely detectable level until ingestion by a new host causes excystation to occur. This period of suspended animation may last for several months – or even years – and appears to depend on the amylopectin reserves of the parasite (Vetterling & Doran, 1969). Just how does ingestion by a suitable host reactivate the dormant parasite, and just how does the sporozoite make its escape from the protective environment of the sporocyst within the oocyst?

Oocysts of many species of coccidia have a prominent micropyle, and sporocysts of most species have a Stieda body; it is through the gap left by the removal of the Stieda body and through the micropyle that the sporozoite makes its escape. What is the signal though which causes the sporozoite to make this escape, and how is the Stieda body removed and the micropyle made penetrable? Most authorities agree that CO_2 provides the primary stimulus in the case of mammalian coccidia, and under experimental conditions excystation can be achieved by exposure to CO_2 followed by digestive enzymes. Carbon dioxide is undoubtedly present in the intestine, particularly in the case of ruminants. Figures are, however, hard to come by, and a useful piece of research would be to measure CO_2 tensions in the intestine of species and regions of the intestine relevant to studies with the coccidia; we have little idea of the relevance of the CO_2 tensions used in laboratory experiments to the conditions prevailing *in vivo*. Trypsin plus bile salts bring about excystation *in vitro* following exposure to CO_2, but not without this prior exposure.

Incubation at body temperature is required for optimal enzymic activity, and the small amounts of excystation which is sometimes noted on prolonged incubation in the absence of extraneous CO_2 could well be due to formation of respiratory CO_2 which could then trigger excystation. During this secondary phase of excystation sporozoites start to move actively within the sporocyst, the Stieda body swells and then disappears, and the sporozoites escape through the resulting hole in the sporocyst into the oocyst fluid, and subsequently penetrate the micropyle, thus leaving the oocyst. Without the prior CO_2 treatment no activation of the sporozoites and no escape takes place. Temperature alone is insufficient to cause sporozoite activation. Bile may well be the activating factor as has been shown for merozoites by Speer, Hammond & Kelley (1970). The oocyst is permeable to CO_2 – CO_2 escapes through the wall as a by-product of respiration during the sporulation phase – but not to much larger molecules. Carbon dioxide in some way alters the oocyst wall to make it permeable to larger molecules such as trypsin and bile, which can then penetrate, digest the proteinaceous Stieda body and activate the sporozoite. Carbon dioxide only has this effect when incubation is carried out at body temperature – suggesting that it is an enzymic process rather than a physico-chemical one. When normal oocysts are incubated with CO_2, no alteration is immediately obvious; they just become susceptible to the action of trypsin and bile. Working with *E. magna* from the rabbit, we made two interesting observations (Ryley & Robinson, 1976). *E. magna* is characterized by a particularly prominent micropyle – a sort of frilled collar round the anterior pole of the oocyst. When oocysts were treated with hypochlorite – we used 10 % hypochlorite for 20 min – the oocysts lost their characteristic orange-brown colour, and lost their micropyle as well. The wall was smooth and refractile all over and the viability of the contents was in no way impaired. Just the colour and the micropyle had gone. As previously mentioned, Nyberg & Knapp (1970*b*) had shown with *E. tenella* and the electron microscope that hypochlorite removed the outer layer of the oocyst wall, and we had shown that this represented approximately 20 % of the wall on a dry weight basis (Ryley, 1973). We therefore believe that with *E. magna*, the colour and the micropyle are features of this outer layer. True, the colour could have been destroyed by the bleaching action of the hypochlorite. The micropyle we assume is a break in the outer layer, and that the characteristic collar is merely the configuration of the edges of the ectocyst around the apical hole in it.

The other feature of the oocyst following hypochlorite treatment we found was that on incubation in the presence of CO_2, the wall lost its normal refractility and indeed its turgor, and assumed a greyish, non-refractile and often irregular aspect. The outer hypochlorite-soluble layer appeared to maintain wall rigidity and mask the optical changes brought about by CO_2. On treatment with trypsin plus bile, excystation took place, and sporozoite escape was possible not just through the micropyle – not that there was any longer a micropyle present – but through any part of the oocyst wall. Carbon dioxide had, therefore, done something to the inner oocyst wall to render it permeable to digestive enzymes and to make it penetrable by sporozoites. The effect requires body temperature and also requires that the oocyst be sporulated; unsporulated oocysts have walls and micropyles which in no way look different to sporulated ones, and yet they are not affected by incubation with CO_2. We have suggested that CO_2 triggers the release or unmasking of an enzyme already present in the sporulated oocyst (but not the unsporulated one) which attacks the oocyst wall from the inside to bring about the necessary changes observed. What this enzyme is or where it is located in the oocyst we have no idea. It could be that the large amounts of leucine amino-peptidase observed by Wang & Stotish (1978) are in some way implicated, although their observations were made with *E. tenella*, which is not so relevant in this context. The precise mechanism of operation of the CO_2 stimulus would repay careful investigation.

The coccidia of chickens do not appear to have micropyles in their natural state and CO_2 does not appear to play a significant part in the excystation process. Oocysts of the chicken coccidia do, however, become susceptible to the excysting activity of trypsin plus bile following *prolonged* incubation with *high* concentrations of CO_2, and Nyberg & Knapp (1970*a*) described a 'micropylar' alteration in the apical part of the wall of *E. tenella* oocysts following CO_2 treatment. There is, however, a very marked difference in susceptibility to CO_2 pre-treatment between the avian and mammalian coccidia. In searching for some enzymic mechanism for the effect of CO_2, assays of potentially important enzymes in the two types of oocyst would be profitable. In the case of the poultry coccidia, it is the shearing and grinding forces in the gizzard that appear to be important in providing access of trypsin and bile to the sporocyst. Although so resistant to chemical attack, the oocyst wall does appear susceptible to mechanical forces. It would be useful to have a quantitation of the mechanical properties and susceptibilities of the wall.

Following the primary stimulus, be it mechanical or CO_2, exposure to trypsin and bile causes the sporozoites to become motile within the sporocyst, and following dissolution of the Stieda body, to escape through the resulting hole. This apical pore in the sporocyst is of a smaller diameter than that of the sporozoite, and during its escape, a constriction passes down its body. With no external organs of locomotion, the question arises how does the sporozoite force itself through this tiny hole? It may be that as it protrudes its tip through the hole and the pellicle makes contact with the sporocyst wall a constriction is established, and it is this constriction which makes purchase on the sporocyst aperture, and by movement of the sub-pellicular microtubules, and thus the constriction along the body of the sporozoite, allows the sporozoite to escape. It is possible, however, that sporozoite escape is passive. It could be that the sporocyst fluid is osmotically more concentrated than the external environment, and that when the Stieda body becomes permeable or lost, influx of fluid into the sporocyst causes expulsion of the sporozoites. That this latter suggestion is a real possibility may be surmised from the observation that sporocyst residual bodies occasionally escape from the sporocyst intact. The residual body appears to be a membrane-bound collection of assorted organelles and particles (Ryley, Bentley, Manners & Stark, 1969), with no conceivable means of active locomotion. During excystation it usually disintegrates within the sporocyst into a collection of tiny particles which can be seen by phase-contrast microscopy to move violently by Brownian motion.

Occasionally, however, the residual body does not disintegrate, and although it may remain within the otherwise empty sporocyst case, I have occasionally seen it squeezing its way through the apical pore with a constriction just like that of an escaping sporozoite. A study of the osmotic relationships between the sporocyst, oocyst and intestinal fluids could be very relevant to the mechanisms of sporozoite liberation.

Homing and cell invasion

In the *in vivo* situation sporozoites are liberated somewhere along the intestinal tract. Unless they respond actively, they will be carried passively with the digesting gut contents along the intestine and eventually voided in the faeces. The coccidia are remarkably species specific for development – although excystation can take place in an unsuitable host – and within that species, show additionally a strict site specificity. *In vitro* sporozoites can be seen under suitable conditions to flex, gyrate or glide. Movement is possibly dependent on bodily contortions brought about by means of the sub-pellicular microtubules, and the mechanism and nature of motility could form a very useful topic for study; a preliminary report, based on analysis of ciné film, of the movement of *E. acervulina* sporozoites was given by Doran, Jahn & Rinaldi (1962). An alternative and most interesting possibility, however, is that movement may be associated with some form of sporozoite secretion, or something akin to the 'camphor boat'. The anterior third of the sporozoite, with its massive array of rhoptries and micronemes, has the appearance of a secretory complex, and I have noted globules apparently emanating from the posterior tip of sporozoites subsequent to movement. Vanderberg (1974) has made similar observations of droplets leaving the trailing ends of gliding sporozoites of *Plasmodium* and has marshalled arguments to suggest that sporozoite motility depends on secretions produced by the anterior rhoptry–microneme complex. Conditions for movement *in vitro*, on say a glass slide, are different from conditions *in vivo* through a pulsating, particulate viscous suspension of gut contents. Is it purely by chance that sporozoites come up against epithelial cells in the particular part of the intestine specific for that organism – or is there some form of chemotactic navigation? If so, are particular chemicals involved, or do coccidia move along specific gradients of pH, oxygen or CO_2 tension, etc? It used to be thought that site specificity was correlated with the rapidity of excystation – since *in vitro E. acervulina* will excyst more rapidly than say *E. tenella* – and that sporozoites were passively liberated in sites suitable for their further development. However, if sporozoites are introduced experimentally into unnatural parts of the intestine – or even the body – some of them at least are able to migrate against the tide to their preferred place of development. It would be a difficult task to sort out possible mechanisms by which coccidia reach their preferred site of development – but nevertheless it is a very important, and as yet unsolved, question.

Ciné films, prepared by the time-lapse technique, of sporozoites in tissue culture are most impressive. Sporozoites are seen darting about, entering cells of the cultured monolayer, leaving them, sometimes passing straight through, and even apparently on occasions penetrating the nucleus. There has been considerable debate as to whether sporozoites do in fact penetrate the host cell, or whether they rather cause an invagination and stretching of the host cell membrane so that the parasite ends up in a vacuole bounded by an inverted host cell membrane. Electron microscopy of tissue-culture cells being invaded by sporozoites, using ruthenium red staining during OsO_4 fixation, has suggested that this latter is most probable (Jensen & Hammond, 1975; Jensen & Edgar, 1978). The speed of invasion is remarkable and the mechanism obscure. Extrusion of the conoid seems to be involved and one can hardly ignore a possible involvement of the rhoptry–microneme complex. Indeed, in studies on the invasion of cultured cells by *E. magna*, Jensen & Edgar (1976) presented electron micrographs apparently showing drops of secretion in the vicinity of the penetrating parasite. Invasion is so rapid that one can hardly imagine it possible that digestive enzymes are secreted to etch a hole through the host cell membrane. Possibly a sporozoite secretion is produced which alters the physico-chemical properties of the membrane and allows the stretching to take place, which is necessary if the membrane is to invaginate and surround the invading parasite. Does invasion thus depend on an active response by the host cell to a passive parasite, or is it an active process initiated and carried through by the invading parasite? Is extrusion of the conoid associated with secretory activity at the crucial moment, or is a rapid and physical 'punching' process necessary for cell invasion?

BIOCHEMICAL BASICS

The biochemistry of coccidia has been reviewed by myself (Ryley, 1973), and more recently by Wang (1978). Conventional biochemistry requires the provision of adequate amounts of pure, uncontaminated parasite material, and this has hitherto meant that studies have been restricted to oocysts and to sporozoites derived therefrom. This is unfortunate since the oocyst is basically a dormant stage, and the activities of many enzymes of interest would be expected to be low. Thus Wang (1978) was unable to detect thymidylate synthetase in crude extracts of unsporulated oocysts of *E. tenella*, although we have been able to detect and characterize the enzyme using more sensitive techniques (Coles, Swoboda & Ryley, 1979). Interesting investigations have been carried out on the biochemical basis of sporulation and of catabolic systems operational in the sporozoite, but it is the growing intracellular parasite which is potentially of prime interest. Histochemical techniques have been tried with intracellular forms, but investigations have tended to favour the easiest recipe – e.g. acid and alkaline phosphatases – rather than the most interesting and relevant enzymes. Thus, little of interest has been revealed other than the possibility of alternation of metabolic pathways during the life-cycle, a situation reminiscent of that found with the African pathogenic trypanosomes. Two breakthroughs in technique have recently been reported which open up interesting opportunities for significant biochemical investigation, and these are discussed below.

Rhoptry–microneme function

Electron microscopical investigations of the coccidia over the last 20 years or so have produced a plethora of pictures illustrating in amazing detail the various stages of the life-cycles of a whole range of species. There has been an inordinate amount of discussion concerning the terminology and phylogeny of the various sub-cellular organelles illustrated, but a marked reticence in defining function. One of the most impressive features of many coccidial zoites – e.g. endozoites of *Sarcocystis tenella* (Ludvík, 1958), sporozoites of *E. tenella* (Ryley, 1969) – is the rhoptry–microneme complex occupying the anterior third of the cell. This complex has all the appearances of a secretory apparatus, and its relative size underlines its potential importance. Two possible functions have been mentioned previously in this review – facilitation of cell invasion and/or parasite motility – but little has been done to substantiate these ideas. De Souza & Souto-Padrón (1978) have shown by the use of ethanolic phosphotungstic acid and the electron micro-scope that the rhoptries and micronemes of *Toxoplasma gondii* are particularly rich in basic proteins. This they suggest indicates their involvement in host-cell penetration by analogy with the work of Kilejian with *Plasmodium lophurae*. She had isolated cytoplasmic granules from erythrocytic stages of *P. lophurae* and shown that they consisted mainly of an unusual protein characterized by a 73% content of histidine (Kilejian, 1974). Further studies (Kilejian, 1976) sug-gested that much of this basic protein was associated with rhoptries and micro-nemes in the polar regions of the parasite, and isolated protein was shown to cause agglutination of erythrocytes and increase their osmotic fragility – a finding which she suggested implied a function in penetration of merozoites into red cells. Using endozoites of *S. tenella*, Dubremetz & Porchet (1978) have shown that when invasion of tissue-culture cells takes place, a variety of materials, apparently secreted by the penetrating parasite, can be detected in the anterior zone of the invaginating host cell membrane – findings in agreement with the studies of Jensen & Edgar (1976) previously mentioned. Dubremetz & Dissous (1978) have developed optimal conditions for disrupting the membranes of *S. tenella* endozoites by means of a French pressure cell without destroying the internal organelles, and have been able by sucrose-gradient centrifugation to separate out the micronemes. If this cell fractionation work can be extended to other species, gives adequately pure preparations and can be performed on a large enough scale, it should make possible an investigation of the nature of the secretory material contained in the micronemes and hopefully give some clues as to its primary function. To me, this is the most interesting challenge which faces the coccidial biochemist today.

Schizont isolation

When considering anabolic rather than catabolic capabilities in the coccidia, the actively growing schizont is the stage of greatest potential interest. In studies already mentioned, Fernando & Pasternak (1977*a*) described the isolation of chick intestinal cells infected with second-generation schizonts of *E. necatrix*. Their interest lay in the nucleus and membranes of the infected host cell rather than in the parasite itself, and parasite material was lost in the discarded debris following host cell rupture (Fernando & Pasternak, 1977*b*). In our laboratories James (1980*a*) has adapted the method of Fernando & Pasternak to isolate pure preparations of immature schizonts of *E. necatrix* and *E. tenella*. Such preparations are metabolically active as shown by their ability to metabolize radioactive glucose. We have been somewhat interested in the thymidylate synthesis cycle from the point of view of chemotherapeutic intervention, and James has been able to extract and characterize serine hydroxymethyl transferase from the isolated parasites. A number of anticoccidial drugs are believed to interfere with nutrient uptake or the ionic status of the coccidia, but limited studies carried out to support these ideas have utilized irrelevant tissues. James (1980*b*) has investigated the uptake of thiamine by isolated schizonts and isolated host intestinal cells and shown that the vitamin is taken up by both active and passive processes. The kinetic constants of the active process were different for parasite and host, and both systems were competitively inhibited by amprolium. The anticoccidial drugs of current significance are the ionophores, particularly monensin and lasalocid. Numerous studies have been carried out on ion translocation in rat liver mitochondria, chick erythrocytes (Wang, 1978) etc, but none with the coccidia themselves. It has been suggested that the ionophores do not have a direct anticoccidial effect, but rather alter the internal pH of the potential host cell, making it an unsuitable environment for parasite development. James started to investigate the effects of monensin and lasalocid on the ionic status of isolated schizonts and isolated intestinal cells, but the work unfortunately had to be terminated before any definite conclusion was reached. Although the amount of pure schizont material which can be isolated in a single preparation is limited, it is to be hoped that this technique will allow the mode of action of the ionophores and other aspects of parasite metabolism to be further investigated.

SEX IN COCCIDIA AND ITS RESEARCH IMPLICATIONS

The terminal stage of the parasitic phase of the coccidian life-cycle is sexual, resulting in the production of the oocyst by the fusion of a micro- and macrogamete. The sexual stages in *E. tenella* develop from second-generation merozoites and there has been much speculation as to where in the life-cycle sexuality is determined. Klimes, Rootes & Tanielian (1972) grew *E. tenella* in tissue culture and claimed that second-generation schizonts were sexually differentiated. Merozoites destined to develop into female gametocytes were strongly PAS-positive, while merozoites which would develop into male gametocytes showed little PAS staining; all merozoites in a schizont were of a similar type. Although there has been no such demonstration of sexual segregation at the first-generation schizogony stage, its possibility is suggested by the work of McDougald & Jeffers (1976); by repeatedly selecting for precocious development, they have obtained a strain of *E. tenella* which produces viable oocysts after only one generation of schizogony. It has long been known that oocysts are bisexual, since infections in chickens can be established from single oocysts. Lee, Remmler & Fernando (1977) showed that sporocysts of *E. tenella* are not sexually differentiated, since they were able to establish infections in chickens with a single sporocyst; it is interesting to note that only 10–15 % of oocysts from these infections would sporulate compared with 70–80 % in their normal infections. Likewise, Lee *et al.* (1977) and Shirley & Millard (1976) were able to establish infections in chick embryos by inoculation of a single sporozoite of *E. tenella*. How and where sexual determination take place is, therefore, still a mystery, but the fact that infections leading to the production of oocysts which will sporulate can be established from a single sporozoite has a practical application in that it means that true clones of coccidia can now be established.

Although experimentation is difficult in view of the dangers of cross-contamination and the necessity for working with extreme caution and in isolators (see final section), recent developments have indicated that genetic studies with the coccidia are now possible. Jeffers (1974) was the first to show that anticoccidial drug resistance could be used as a genetic marker; strains of *E. tenella* resistant to amprolium and decoquinate were crossed to produce doubly-resistant organisms. Joyner & Norton (1975, 1977) described similar experiments with *E. maxima* and presented evidence to support the view that resistance factors were transferred at zygote formation. Precocious development has been used as an alternative genetic marker by Jeffers (1976) and isoenzymes by Shirley (1978). We have put the idea to practical use in carrying out genetic recombination experiments using drug resistance as markers as our contribution to the abortive *E. acervulina*/*E. mivati* controversy (Ryley & Hardman, 1978). It is to be hoped that those with sufficient patience and insight will really open up the field of genetic studies with the coccidia.

LET'S BE PRACTICAL

The coccidia are a fascinating group of organisms and many interesting problems concerning structure, life-cycle, biochemistry, etc. are on hand for the academic seeking pastures new. It must be remembered, however, that the coccidia are parasites, and in the case of poultry coccidia – and to a lesser extent ruminant coccidia – parasites of extreme economic importance. The modern broiler industry could not exist were practical control of coccidiosis not a possibility, and it is likely that coccidiosis is a cause of loss of production, if not of life, in modern intensive systems for raising calves and lambs. The resilience of the oocyst has already been mentioned; coccidia will be found wherever there are susceptible birds and animals, and it seems unlikely that practical measures to remove the organisms from the environment can be devised. There have been sporadic attempts to breed strains of chicken resistant to coccidia – but resistance to coccidiosis does not seem to go hand-in-hand with all the other genetic traits such as body conformation, feed efficiency, etc. so important in poultry breeding. The two most practical approaches are chemoprophylaxis and immunological control; the former, currently, is by far the most important.

Chemotherapy

Chemotherapy of chicken coccidiosis has been practised since the introduction of the sulphonamides in the early 1940's. Although *treatment* of outbreaks of coccidiosis in otherwise unmedicated adult birds is sometimes necessary, the greatest use of drugs is in the *prevention* of the disease in broilers by the administration of an anticoccidial in the feed throughout the 6–8 weeks life of the bird. Over these last four decades, a succession of drugs has been introduced. Early drugs were somewhat limited in their spectrum of activity, and with the need to control at least 6 different species, broader control has been achieved either by the use of mixtures of drugs or by the discovery of new compounds with an inbuilt wider spectrum of activity. Until recently, new drugs were mostly characterized by a greater intrinsic degree of activity and a greater freedom from toxicity than predecessors; it is interesting that with more recent additions to the drug armamentarium – monensin, halofuginone and arprinocid – the margin between incorporation rates necessary to control coccidiosis and those causing a depression in weight gain is remarkably narrow. It is extremely fortunate that there has been a continual succession of new anticoccidial drugs, since one by one, older drugs have fallen foul of the problem of drug resistance; with some drugs resistance has become a problem very quickly, with others it has taken time, but in almost all cases it has eventually emerged. We know almost nothing about the biochemical nature of anticoccidial drug resistance or why and how it emerges. Likewise, there seems no solution, other than to change to a new and unrelated drug, when breakdown occurs. In an attempt to prevent or delay the emergence of resistance, the idea of drug rotation has been introduced; two or three unrelated drugs are used sequentially, the change from one drug to another being made either when the diet is changed during the rearing of a crop of broilers, or alternatively between crops. There is no doubt that this practice prolongs the apparent life of a drug, but there is little evidence to suggest that it prolongs it in relation to the actual cumulative period during which the drug is used. Although the proportion of resistant coccidia in a population may decrease in the absence of drug pressure, there is little evidence to suggest that resistance is actually lost; re-introduction of drug pressure results in a quick reselection for resistance – with the consequent breakdown of prophylaxis. An alternative approach could be the use of a mixture (preferably synergistic) of two or three drugs, such as has been used so successfully in tuberculosis, in the hope of slowing down the emergence of resistance. The problem is to discover two or three such new drugs at the same time to mix together! Drug resistance is one of the most important practical problems requiring attention in the coccidiosis field today, but significant ideas are sparse! The whole topic of resistance has been fully reviewed elsewhere (Ryley, 1980*a*).

Prophylaxis of coccidiosis today relies to a very large extent on the ionophore monensin. This is a most interesting drug in that it has performed very well in the field, but very poorly in the majority of laboratories. In the laboratory where the anticoccidial activity of this drug was discovered, there was a lapse of several years before the drug was evaluated in the field; had it been discovered elsewhere, it is quite likely that it would never have got beyond the laboratory! Anticoccidial activity has been found in an appreciable number of other ionophores; among these lasalocid is available commercially and salinomycin and narasin have been evaluated in the field. Some of the older anticoccidial drugs are still being used – in an attempt to save money, while at the same time keeping some measure of coccidiosis control. In view of the prevalence of resistant strains, it is unlikely that the broiler industry could rely for long on these older drugs alone. Two new synthetic compounds are in the news. Halofuginone is available commercially, and has the distinction of being weight-for-weight the most active compound on the market, with an inclusion rate of 3 p.p.m. Arprinocid has undergone field evaluation and will presumably be available shortly. This latter is an interesting compound in that metabolism to the N-oxide seems to be necessary for it to become active (Latter & Wilson, 1979).

Drug resistance to halofuginone can be readily produced in the laboratory, but this is not the case with monensin. There is no doubt that some strains of some species of coccidia respond less well to monensin than others, but to what extent this is a reflection of a natural variation or to what extent an indication of increasing resistance consequent on exposure to drug is not clear. Chapman (1979) has recently produced evidence to suggest that in the U.K., field exposure of *E. maxima* to monensin is resulting in the emergence of less sensitive strains. A question of particular current interest is if and when resistance to monensin becomes a problem, how will such strains respond to other ionophores? Evidence on this point is limited and conflicting, but I believe cross-resistance would be a problem (Ryley, 1980a). An answer to this question is vital to those considering the development of other ionophores.

It is not easy to discover a new anticoccidial; 20 years work in our own laboratories produced only one compound, 'Statyl'* (methyl benzoquate). Methods for the detection and evaluation of anticoccidial activity are described elsewhere (Ryley, 1980b). Because anticoccidials are used prophylactically rather than therapeutically, it is very important that the drugs are free from toxicity, not only to the growing birds who receive them throughout life, but also to those who may eat the flesh or eggs of treated birds and thus ingest drug residues. It is right and proper that assurances on safety should be obtained before a new anticoccidial is marketed. However, the cost and effort of producing such evidence to satisfy all the regulatory authorities is now becoming so great – outweighing by far the cost and effort involved in discovering the drug in the first place – that anticoccidial research is becoming less and less attractive to the pharmaceutical industry. Such considerations have contributed to our own recent decision to withdraw from active research. The broiler industry is completely dependent on the availability of adequate drugs. The current situation is admittedly satisfactory, but if past experience is anything to go by, there will always be a need for new drugs – and with the economic strictures being placed on anticoccidial research, there is no guarantee that an adequate future supply of drugs will be forthcoming.

* 'Statyl' is a trade mark, the property of Imperial Chemical Industries Limited.

Immunological control

Immunology of the coccidia is an extremely complicated topic; immunity is highly species specific, and to some extent may be strain specific. The field has recently been reviewed by Rose (1978). In view of the uncertainties of anti-coccidial drug discovery and the ever increasing economic pressures in poultry production and the cost of prophylactic medication, some form of coccidiosis vaccination could be an extremely attractive proposition. For the development of immunity it seems that it is necessary for the parasite to develop within the host; there is no evidence to date which suggests that it can be generated in response to the administration of non-viable parasite-derived material. Two approaches to the problem are currently being made: controlled exposure to virulent coccidia, or the development of attenuated strains for use in a vaccine. Coccivac, a mixture of live oocysts of normal virulent strains of 8 different species, is available in the USA, and is administered to young birds, sometimes under cover of a second-rate anticoccidial to prevent the infection getting out of hand, but without completely stopping coccidial reproduction. It has not been possible to get a reliable idea of how widely used this product is. Some multiplication of the parasite is necessary to allow immunity to develop, and continual re-infection is probably required to sustain an adequate level of immunity. The use of a poly-valent vaccine such as Coccivac is not ideal, since species may be introduced into a flock which were not previously present, and the balance between sub-clinical and clinical coccidiosis is extremely delicate. Because in the case of broilers it is important that birds put on the maximum amount of weight with the minimum amount of feed in the shortest possible time, and no setback in growth, even temporary, can be tolerated, the Coccivac type of product is potentially more useful in laying birds and broiler breeders. It is interesting that Joyner & Norton (1973, 1976) have shown that daily administration of 1–20 oocysts of *E. acervulina*, *E. maxima* or *E. tenella* for a period of 3–4 weeks could produce a substantial degree of immunity, presumably without any detectable clinical disease; the reliable administration of such small numbers of oocysts daily could present problems!

An alternative approach to immunization has been to attenuate initially virulent strains, either by repeated passage through chick embryos (Long, 1974) or by selection for precociousness (Jeffers, 1975). The former method has only so far worked with *E. tenella* and *E. mivati*, and the latter only with *E. tenella*. Attenuation is associated with alterations to the second generation of schizogony in the case of *E. tenella* and a consequent reduction in pathogenicity; schizonts are smaller, produce fewer merozoites and are located superficially rather than in the deeper tissues. Jeffer's most attenuated strain in fact doesn't have a second genera-tion of schizogony at all, but proceeds directly from the first asexual generation to gametogony (McDougald & Jeffers, 1976). As a result of these modifications in the life-cycle the reproductive potential of the attenuated strain is reduced. Floor pen trials have been carried out to compare the immunogenicity of Jeffer's strains of *E. tenella* with that present in Coccivac (Johnson, Reid & Jeffers, 1979). It is interesting that his most attenuated strain – the one without any second-generation schizogony – didn't produce any immunity, whereas a less attenuated strain was satisfactory. Whether it will be possible to attenuate all the species which need to be present in a polyvalent vaccine remains to be seen. If such a vaccine were to be widely used, there would be considerable production problems to provide the astronomical numbers of oocysts which would be required.

A WORD OF WARNING; AN APPEAL FOR CAREFUL WORK

It seems to me likely that some of the coccidiosis literature is unreliable because of the use of impure strains, particularly in studies on life-cycles, immunology and drug-resistance. An interesting situation with the Houghton strain of *E. tenella* was uncovered by Chapman (1975). He was able to obtain drug resistance to quinolone anticoccidials after a single experimental passage – a finding which did not agree with previous observations in other laboratories. Chapman was able to demonstrate that the Houghton strain had unwittingly become contaminated with drug-resistant organisms during routine passage sometime subsequent to 1969 (when an uncontaminated sample had been frozen in liquid nitrogen) – but of course this was not obvious until specific drug pressure was applied. Our own experience had been entirely similar; during the development of 'Statyl' some 12 years ago, it was problems of cross-contamination between drug-sensitive and drug-resistant strains which caused us to design and build extremely expensive facilities specifically for coccidiosis research.

E. mivati was described and named by Edgar & Seibold in 1964. Long & Tanielian described its isolation from the Lebanon in 1965, and for a while it was regarded as a valid species. However, in 1973 Long published a paper in which he compared the abilities of two strains each of *E. acervulina* and *E. mivati* to develop in unusual sites, and reported on a number of cross-immunity experiments. As a result of his experiments he suggested that *E. mivati* should be regarded as a variant of *E. acervulina* rather than a distinct species. This sub-specific nature of *E. mivati* was maintained for several years in a series of publications (reviewed by Joyner & Norton, 1980), and other laboratories became involved. Shirley (1979) then demonstrated that the strains of *E. mivati* maintained at Houghton were in fact contaminated with *E. acervulina*, and that the degree of contamination had increased with the passage of time – a situation which could have been anticipated from some of Long's remarks on oocyst size and shape in his 1973 paper. A confusing situation had arisen and a great deal of time, effort and argument had been expended because impure cultures of coccidia had unwittingly been used in the first place.

In our own contribution to the *E. acervulina*/*E. mivati* controversy (Ryley & Hardman, 1978), although we took the precaution of carrying out all our work in isolators, we made the mistake of assuming that the 'pure' cultures of these organisms initially supplied by Dr S. A. Edgar were in fact pure; measurements of size and shape did not lead us to suspect otherwise, but Joyner (personal communication) has subsequently examined samples of the original cultures supplied by Edgar, and believes that the *E. mivati* was in fact contaminated with *E.*

acervulina. This no doubt explains the immunological cross-reactions we observed in our studies – a situation analogous to that of Long (1973). Fortunately for us, the experiments showing that the two organisms will not inter-breed are wholly reliable; due to the fact that *E. acervulina* is basically more sensitive to methyl benzoquate than *E. mivati*, repeated passage of Edgar's *E. mivati* in the presence of drug led to the production of a *pure* drug-resistant line of *E. mivati*, which would not inter-breed with resistant lines of *E. acervulina.*

It is essential that when work of a critical nature is undertaken, where the outcome depends on the purity and integrity of the organism used, that cultures are started from single oocyst isolates and experiments are carried out with birds raised and maintained in isolators. If adequate facilities are not available, it would be better that the work were not done, lest further confusion be added to the coccidiosis literature. Short-cuts, such as relying on the integrity of other peoples' cultures, should not be taken.

A further practical issue has been raised by the *E. acervulina/E. mivati* controversy, and that is the status of *E. hagani*, and particularly *E. mitis*. Published descriptions, in view of the known variability of characters in the coccidia, make it difficult to distinguish between *E. mitis* and *E. mivati*. If the two could be shown to be identical, the name *E. mitis* would have priority. The problem is, however, that the strains of *E. mitis* (and *E. hagani*) on which the literature descriptions were based are no longer available, and the problems cannot, therefore, be resolved experimentally. Coccidia can be readily preserved in liquid nitrogen. Work with coccidia should only be done with strains which are freely available between workers and samples of which have been preserved in liquid nitrogen. When new strains are isolated, and particularly when new species are described, work should not be published until sufficient material has been produced to make depositions in the liquid nitrogen banks of several laboratories – until such time as a central reference collection can be established. Editors of journals usually require sufficient experimental detail to be given to allow work to be repeated, but work of this nature cannot be repeated unless the organism concerned is also freely available.

CONCLUSIONS

The field of coccidiosis is full of problems to provoke the curiosity of the scientist however academic or applied his outlook may be. The solution of some of them will help maintain the viability of the poultry industry and be a real contribution to world nutrition, the pursuit of all will be a challenge to the scientist as he studies this most fascinating group of organisms.

REFERENCES

CHAPMAN, H. D. (1975). *Eimeria tenella* in chickens: development of resistance to quinolone anticoccidial drugs. *Parasitology* **71**, 41–9.

CHAPMAN, H. D. (1979). Studies on the sensitivity of recent field isolates of *E. maxima* to monensin. *Avian Pathology* **8**, 181–6.

CLARKE, P. L. (1978a). The effect of *Eimeria tenella* on the frequency of caecal evacuation in the domestic fowl. *Parasitology* **77**, vii.

CLARKE, P. L. (1978*b*). The structure of the ileo-caeco-colic junction of the domestic fowl (*Gallus gallus* L.). *British Poultry Science* **19**, 595–600.

CLARKE, P. L. & CROMPTON, D. W. T. (1978). The avian caecum as an environment for parasites. *Parasitology* **77**, xxii.

COLES, A. M., SWOBODA, B. E. P. & RYLEY, J. F. (1979). Thymidylate synthetase as a chemotherapeutic target in the treatment of avian coccidiosis. *Parasitology* **79**, li.

CROMPTON, D. W. T. & NESHEIM, M. C. (1976). Host–parasite relationships in the alimentary tract of domestic birds. *Advances in Parasitology* **14**, 95–194.

DE SOUZA, W. & SOUTO-PADRÓN, T. (1978). Ultrastructural localization of basic proteins on the conoid, rhoptries and micronemes of *Toxoplasma gondii*. *Zeitschrift für Parasitenkunde* **56**, 123–9.

DORAN, D. J., JAHN, T. L. & RINALDI, R. (1962). Excystation and locomotion of *Eimeria acervulina* sporozoites (motion picture). *Journal of Parasitology* **48**, Suppl. p. 32. Abstr. no. 57.

DUBREMETZ, J. F. & DISSOUS, C. (1978). Subcellular fractionation of *Sarcocystis tenella* endozoites. *Abstracts 4th International Congress of Parasitology, Warszawa*. F, 46–7.

DUBREMETZ, J. F. & PORCHET, E. (1978). *Sarcocystis tenella*: pénétration des endozites dans la cellule hôte *in vitro*. *Journal of Protozoology* **25**, 52A. Abstr. no. 159.

EDGAR, S. A. & SEIBOLD, C. T. (1964). A new coccidium of chickens, *Eimeria mivati* sp.n. (Protozoa: Eimeriidae) with details of its life history. *Journal of Parasitology* **50**, 193–204.

FERNANDO, M. A. & PASTERNAK, J. (1977*a*). Isolation of chick intestinal cells infected with second-generation schizonts of *Eimeria necatrix*. *Parasitology* **74**, 19–26.

FERNANDO, M. A. & PASTERNAK, J. (1977*b*). Isolation of host-cell nuclei from chick intestinal cells infected with second-generation schizonts of *Eimeria necatrix*. *Parasitology* **74**, 27–32.

FERNANDO, M. A., PASTERNAK, J., BARRELL, R. & STOCKDALE, P. H. G. (1974). Induction of host nuclear DNA synthesis in coccidia-infected chicken intestinal cells. *International Journal for Parasitology* **4**, 267–76.

IKEDA, M. (1957). Factors necessary for *E. tenella* infection of the chicken. V. Organ specificity of *E. tenella* infection. *Japanese Journal of Veterinary Science* **19**, 105–14.

JAMES, S. (1980*a*). Isolation of second-generation schizonts of avian coccidia and their use in biochemical investigations. *Parasitology* **80** (in the Press).

JAMES, S. (1980*b*). Thiamine uptake in isolated schizonts of *Eimeria tenella* and the inhibitory effect of amprolium. *Parasitology* **80** (in the Press).

JEFFERS, T. K. (1974). Genetic transfer of anticoccidial drug resistance in *Eimeria tenella*. *Journal of Parasitology* **60**, 900–4.

JEFFERS, T. K. (1975). Attenuation of *Eimeria tenella* through selection for precociousness. *Journal of Parasitology* **61**, 1083–90.

JEFFERS, T. K. (1976). Genetic recombination of precociousness and anticoccidial drug resistance in *Eimeria tenella*. *Zeitschrift für Parasitenkunde* **50**, 251–5.

JENSEN, J. B. & EDGAR, S. A. (1976). Possible secretory function of the rhoptries of *Eimeria magna* during penetration of cultured cells. *Journal of Parasitology* **62**, 988–92.

JENSEN, J. B. & EDGAR, S. A. (1978). Fine structure of penetration of cultured cells by *Isospora canis* sporozoites. *Journal of Protozoology* **25**, 169–73.

JENSEN, J. B. & HAMMOND, D. M. (1975). Ultrastructure of the invasion of *Eimeria magna* sporozoites into cultured cells. *Journal of Protozoology* **22**, 411–15.

JOHNSON, J., REID, W. M. & JEFFERS, T. K. (1979). Practical immunization of chickens against coccidiosis using an attenuated strain of *Eimeria tenella*. *Poultry Science* **58**, 37–41.

JOYNER, L. P. & NORTON, C. C. (1973). The immunity arising from continuous low-level infection with *Eimeria tenella*. *Parasitology* **67**, 333–40.

JOYNER, L. P. & NORTON, C. C. (1975). Transferred drug-resistance in *Eimeria maxima*. *Parasitology* **71**, 385–92.

JOYNER, L. P. & NORTON, C. G. (1976). The immunity arising from continuous low-level infection with *Eimeria maxima* and *Eimeria acervulina*. *Parasitology* **72**, 115–25.

JOYNER, L. P. & NORTON, C. C. (1977). Further observations on the genetic transfer of drug resistance in *Eimeria maxima*. *Parasitology* **74**, 205–13.

JOYNER, L. P. & NORTON, C. C. (1980). The *Eimeria acervulina* complex: problems of differentiation of *Eimeria acervulina*, *E. mitis* and *E. mivati*. *Protozoological Abstracts* **4** (in the Press).

KILEJIAN, A. (1974). A unique histidine-rich polypeptide from the malaria parasite, *Plasmodium lophurae*. *Journal of Biological Chemistry* **249**, 4650–5.

KILEJIAN, A. (1976). Does a histidine-rich protein from *Plasmodium lophurae* have a function in merozoite penetration? *Journal of Protozoology* **23**, 272–7.

KLIMES, B., ROOTES, D. G. & TANIELIAN, Z. (1972). Sexual differentiation of merozoites of *Eimeria tenella*. *Parasitology* **65**, 131–6.

LATTER, V. S. & WILSON, R. G. (1979). Factors influencing the assessment of anticoccidial activity in cell culture. *Parasitology* **79**, 169–75.

LEE, E-H., REMMLER, O. & FERNANDO, M. A. (1977). Sexual differentiation in *Eimeria tenella* (Sporozoa: Coccidia). *Journal of Parasitology* **63**, 155–6.

LONG, P. L. (1973). Studies on the relationship between *Eimeria acervulina* and *Eimeria mivati*. *Parasitology* **67**, 143–55.

LONG, P. L. (1974). Further studies on the pathogenicity and immunogenicity of an embryo-adapted strain of *Eimeria tenella*. *Avian Pathology* **3**, 255–68.

LONG, P. L. & TANIELIAN, Z. (1965). The isolation of *Eimeria mivati* in Lebanon during the course of a survey of *Eimeria* spp. in chickens. '*Magon*' *Institut de recherches agronomiques, Scientific Series* No. 6, 1–18.

LUDVÍK, J. (1958). Elektronenoptische Befunde zur Morphologie der Sarcosporidien (*Sarcocystis tenella* Railliet 1886). *Zentralblatt für Bakteriologie, Parasitenkunde, Infektionskrankheiten und Hygiene Erste Abteilung, Originale*. **172**, 330–50.

McDOUGALD, L. R. & JEFFERS, T. K. (1976). *Eimeria tenella* (Sporozoa: Coccidia): gametogony following a single asexual generation. *Science* **192**, 258–9.

NYBERG, P. A. & KNAPP, S. E. (1970a). Scanning electron microscopy of *Eimeria tenella* oocysts. *Proceedings of the Helminthological Society of Washington* **37**, 29–32.

NYBERG, P. A. & KNAPP, S. E. (1970b). Effect of sodium hypochlorite on the oocyst wall of *Eimeria tenella* as shown by electron microscopy. *Proceedings of the Helminthological Society of Washington* **37**, 32–6.

ROSE, M. E. (1978). Immune responses of chickens to coccidia and coccidiosis. In *Avian Coccidiosis* (ed. P. L. Long, K. N. Boorman and B. M. Freeman), pp. 297–336. Edinburgh: British Poultry Science Ltd.

RYLEY, J. F. (1969). Ultrastructural studies on the sporozoite of *Eimeria tenella*. *Parasitology* **59**, 67–72.

RYLEY, J. F. (1973). Cytochemistry, physiology and biochemistry. In *The Coccidia* (ed. D. M. Hammond and P. L. Long), pp. 145–181. Baltimore: University Park Press.

RYLEY, J. F. (1975). Why and how are coccidia harmful to their host? In *Pathogenic Processes in Parasitic Infections*. Symposia of the British Society for Parasitology, vol. **13** (ed A. E. R. Taylor and R. Muller), pp. 43–58. Oxford: Blackwell.

RYLEY, J. F. (1980a). Drug resistance in coccidia. *Advances in Veterinary Science and Comparative Medicine* **24** (in the Press).

RYLEY, J. F. (1980b). Screening for and evaluation of anticoccidial activity. *Advances in Pharmacology and Chemotherapy* **17** (in the Press).

RYLEY, J. F., BENTLEY, M., MANNERS, D. J. & STARK, J. R. (1969). Amylopectin, the storage polysaccharide of the coccidia *Eimeria brunetti* and *E. tenella*. *Journal of Parasitology* **55**, 839–45.

RYLEY, J. F. & HARDMAN, L. (1978). Speciation studies with *Eimeria acervulina* and *Eimeria mivati*. *Journal of Parasitology* **64**, 878–81.

RYLEY, J. F. & ROBINSON, T. E. (1976). Life cycle studies with *Eimeria magna* Pérard, 1925. *Zeitschrift für Parasitenkunde* **50**, 257–75.

SHIRLEY, M. W. (1978). Electrophoretic variation of enzymes: a further marker for genetic studies of the *Eimeria*. *Zeitschrift für Parasitenkunde* **57**, 83–7.

SHIRLEY, M. W. (1979). A reappraisal of the taxonomic status of *Eimeria mivati* Edgar and Seibold 1964, by enzyme electrophoresis and cross-immunity tests. *Parasitology* **78**, 221–37.

SHIRLEY, M. W. & MILLARD, B. J. (1976). Some observations on the sexual differentiation of *Eimeria tenella* using single sporozoite infections of chicken embryos. *Parasitology* **73**, 337–41.

SPEER, C. A., HAMMOND, D. M. & KELLEY, G. L. (1970). Stimulation of motility in merozoites of five *Eimeria* species by bile salts. *Journal of Parasitology* **56**, 927–9.

STOTISH, R. L., WANG, C. C. & MEYENHOFER, M. (1978). Structure and composition of the oocyst wall of *Eimeria tenella*. *Journal of Parasitology* **64**, 1074–81.

THOMPSON, J. E., FERNANDO, M. A. & PASTERNAK, J. (1979). Induction of gel phase lipid in plasma membrane of chick intestinal cells after coccidial infection. *Biochimica Biophysica Acta.* **555**, 472–84.

VANDERBERG, J. P. (1974). Studies on the motility of *Plasmodium* sporozoites. *Journal of Protozoology* **21**, 527–37.

VETTERLING, J. M. & DORAN, D. J. (1969). Storage polysaccharide in coccidial sporozoites after excystation and penetration of cells. *Journal of Protozoology* **16**, 772–5.

VETTERLING, J. M., MADDEN, P. A. & DITTEMORE, N. S. (1971). Scanning electron microscopy of poultry coccidia after *in vitro* excystation and penetration of cultured cells. *Zeitschrift für Parasitenkunde* **37**, 136–47.

WANG, C. C. (1978). Biochemical and nutritional aspects of coccidia. In *Avian Coccidiosis* (ed. P. L. Long, K. N. Boorman and B. M. Freeman), pp. 135–84. Edinburgh: British Poultry Science Ltd.

WANG, C. C. & STOTISH, R. L. (1978). Multiple leucine aminopeptidases in the oocysts of *Eimeria tenella* and their changes during sporulation. *Comparative Biochemistry and Physiology* **61**B, 307–13.

Parasitology (1980), **80**, 393–412

With 1 *figure in the text*

Invertebrate immunity

ANN M. LACKIE

*Department of Zoology, The University, Glasgow G*12 8*QQ*

(*Accepted* 19 *October* 1979)

INTRODUCTION

One of the main problems which has beset the study of immunity in invertebrates is that these animals have been considered by many workers as mere
evolutionary precursors of the vertebrates. As a consequence, a tremendous amount
of energy has been expended in searching for the progenitors of the T-cell and for
evidence of specific memory. This is unfortunate: although the search for the
origins of the vertebrate immune system is undeniably fascinating, and we can
undoubtedly use our knowledge of vertebrate immunology and its techniques to
great advantage in studying the invertebrates, we still need to keep an open mind
in interpretation of much of the data. There are many areas of invertebrate
immunology where our knowledge is still at the descriptive stage, and we are a
long way from understanding the mechanisms behind the observed response. One
particular barrier to our understanding derives from the corruption of the meaning
of 'immunity', as the word has become inexorably identified with the highly
sensitive and specific response demonstrated by vertebrates. Linked with this is
the subconscious acceptance of the involvement of specific memory and the enhanced secondary response, immunoglobulin and all the other paraphernalia of
the vertebrate system. It is frequently forgotten that immunity may also be
provided by non-specific but nonetheless effective means. This concept of immunity
is easier to understand if we stop worrying about how to explain a system which
can respond to thorns and to *pertussis* vaccine, and return instead to MacFarlane
Burnet's well-worn but gratifyingly apt terms 'self' and 'non-self'. At the simplest
level, then, an animal demonstrates immunity if it has within itself tissues that
are capable of recognizing and protecting the animal against 'non-self'. The
mechanisms for recognition and protection may well vary and become increasingly
complex, culminating in the accelerated specific secondary responses of the homo-
eothermic vertebrates, but as long as the most basic system provides protection
it can be considered as immunity.

One way of attempting to understand invertebrate immune systems is to try to
conceive a system from first principles – and here we already have a problem.
One cannot simply arrange to have 2 interacting mechanisms, the humoral and
cellular components, because there are a large number of invertebrates with a
'solid' body and no circulatory system. Indeed, it is not until the level of organization of the Annelida that animals start to have a system for circulating blood. So,
we need to start with the premise that cellular immunity came first, and to con-

sider its functional requirements. At its most primitive, in an animal whose cells are relatively unorganized into tissues, the immune system may comprise all or most of the cells of the animal, any of which can recognize and respond to 'non-self'; the protective response might require a co-operative effort by neighbouring cells, for example, the production of a substance which kills or walls-off the foreign material. Such a system would provide a localized but effective response, but would also have the disadvantage of interrupting the normal behaviour and function of the responding cells.

The next step in phylogeny involves the specialization of cells into tissues with specific functions, and concomitantly the appearance of a population of cells whose function is to act as itinerant trouble-shooters within the body; these cells – which can now be called leucocytes for convenience – may recognize and respond to wounds as well as to 'non-self'.

In solid-bodied invertebrates, without a body cavity or circulatory system, the effectiveness of the immune response must be dependent on there being a large number of freely mobile leucocytes within the tissues. With the evolution of the body cavity and then, as animals increase in size, a circulatory system, the effective range of the cells of the immune system is greatly increased; because of improved communication, specialization and division of labour amongst leucocytes becomes possible. At this stage, it also becomes feasible to incorporate into the immune system a humoral component, which may act totally independently, but which ideally should interact co-operatively with the cellular components.

It is also at this stage that our problems of interpretation become most acute, for we are now presented with an immune system which appears to have many of the components of the vertebrate immune system – circulating blood; several types of circulating granular and agranular leucocytes, some of which are phagocytic; sessile phagocytes in 'reticular tissue'; perhaps a distinct haemopoietic organ for production of leucocytes; and graft rejection may occur. There is one drawback – there is no evidence for the presence of immunoglobulin in invertebrates; but in spite of this, the temptation to seek homologies with the vertebrate system may be overpowering. There is thus, for example, a regrettable tendency to apply the emotive and suggestive terms – lymphocyte, plasma cell, macrophage – to invertebrate leucocytes, although evidence for analogy or homology with the vertebrate equivalents may be slight.

Rather than seeking for the precursors of vertebrate immune mechanisms, I would suggest that we need to consider the invertebrate immune response in its own right for, as I hope to show, the basic strategies underlying recognition may be rather different. In this context, we need to know considerably more, not only about the specificity of recognition but also about the non-specificity and, in particular, the response to abiotic objects, for this should provide fundamental information about the most basic mechanisms for discriminating between self and non-self.

My aims have been to introduce some of the current ideas about invertebrate immunity and to try to relate them to the biology of invertebrates. I have attempted not only to point out how and where problems in interpretation can arise, but also to indicate areas of the subject where future research should provide us with some much needed answers. My choice of the literature cited has been based on the three criteria – useful reviews, most recent papers, or simply, those papers which I have found most stimulating.

I. MECHANISMS OF THE IMMUNE RESPONSE IN INVERTEBRATES

As the invertebrates are such a diverse group of animals it is obviously impossible, and unnecessary, to go into details here of the manifestations of the immune response; suffice it to say that the observable response *in vivo* is primarily cellular – phagocytosis, encapsulation and occasionally, cytotoxicity responses. Manifestations of the humoral response *in vivo* are less easily observed, but bacteriocidins have been found, a lytic activity similar to that of the 'non-specific' components of the complement system appears to exist in some animals (Day, Gewurz, Johannsen, Finstad & Good, 1970; Aston & Chadwick, 1978) and there are the 'haemagglutinins' whose activity has mainly been investigated *in vitro*. The following section discusses some of the functions of the humoral and cellular components of the immune response and their interactions.

(1) *Cellular immunity – what do the cells do?*

(*a*) Cell types and their functions

The early literature is full of papers describing many categories of invertebrate leucocytes, submerging the reader under complex terminology and elegant histological descriptions. The present tendency is to simplify the classification wherever possible, by attempting to show ontogenetic relationships and to relate structure to function. Some of the types of leucocyte and their functions are listed in Table 1; note that I have used the term 'leucocyte' as a generic term to encompass all types of white cell thought to be reactive in the immune response. 'Haemocytes' and 'coelomocytes' are leucocytes found in animals with a body cavity which is mainly haemocoelic (molluscs, arthropods) or mainly coelomic (annelids, sipunculids, echinoderms), respectively.

Phagocytosis and encapsulation appear to be universal phenomena amongst the invertebrates. The mechanisms of phagocytosis have been especially well studied in insects and molluscs. In insects, phagocytosis is performed most efficiently by the granular plasmatocytes, either on their own, or in combination with granulocytes in nodules formed *in vivo* (Ratcliffe & Gagen, 1976). Acid phosphatase activity is found associated with the phagolysosome and around ingested test particles (Rowley & Ratcliffe, 1979), and the metabolism of haemocytes during phagocytosis has been investigated and discussed by Anderson (1975). Phagocytosis and encapsulation by molluscan haemocytes is accompanied by the release of lysosomal enzymes such as β-glucuronidase and acid phosphatase. (Cheng, 1976; Cheng & Garrabrant, 1977; Sminia, 1972).

Table 1. *Some invertebrate leucocytes and their functions*

Group	Cell types and functions	Phago-cytosis	Encapsu-lation	Trans-plant ation reaction	Useful references
Sponges	Archaeocytes (wandering cells which differentiate into other cell types and can act as phagocytes)	+	+	?*	
Coelenterates	Amoebocytes; 'lymphocytes'	+	.	?	Hildemann *et al.* (1977)
Nemertines	Agranular leucocytes; granular 'macrophage-like' cells	+	.	+	Langlet & Bierne (1977)
Annelids	Basophilic amoebocytes (accumulate as 'brown bodies'); acidophilic granulocytes	+	+	+	Hostetter & Cooper (1972)
Sipunculids	Several types	+	+	+	Cushing & Boraker (1975)
Insects	Several types, depending on Family: e.g. plas-matocytes; granulocytes; spherule cells; coagulo-cytes – blood clotting	+	+	+	Gupta (1979)
Crustaceans	Granular phagocytes; refractile cells which lyse and release contents	+	+	+	Smith & Ratcliffe (1978) Crompton (1967)
Molluscs	Amoebocytes	+	+	+	Tripp (1974), Cheng (1975), Sminia *et al.* (1974)
Echinoderms	Amoebocytes; spherule cells; pigment cells; vibratile cells – blood clotting	+	+	+	Fontaine & Lambert (1977), Bertheussen & Seljelid (1978), Johnson & Chapman (1970a), Reinisch & Bang (1971)
Tunicates	Many types incuding phagocytes; 'lymphocytes'	+	+	+	Smith (1970), Anderson (1971), Freeman (1970)

*? Transplantation reactions occur, but the extent to which the leucocytes are involved is unknown.

If a foreign object is larger than approximately 10 μm in diameter, encapsulation rather than phagocytosis occurs. This response may merely involve 'clumping' of basophilic amoebocytes around damaged tissue as in earthworms (Hostetter & Cooper, 1974) or a more structured capsule may be produced. Good descriptions of encapsulation in molluscs are to be found in papers by Harris (1975) and Sminia, Borghardt-Reinders & van de Linde (1974). Amoebocytes adhere to the object and to each other to build up a capsule several cells thick; the finished capsule may have a fibrous appearance, since the amoebocytes may transform into fibroblast-like cells secreting collagen-like fibres. Encapsulation in insects may take several forms or involve several types of haemocyte, depending on the species of insect investigated. The adhering cells may form 3 distinct layers (Grimstone, Rotheram & Salt, 1967) or may form a less organized capsule with 2 distinct regions (Schmit & Ratcliffe, 1978), or might provide merely a very thin 'sheath' capsule as in certain Diptera which have few circulating haemocytes (Salt, 1970). In addition to phagocytosis or encapsulation, arthropods also have a further defence mechanism – the walling-off of a foreign object by melanin, deposited either by the cells in the capsule (Solangi & Lightner, 1976), or by precipitation from the haemolymph (Poinar & Leutenneger, 1971).

In animals with a circulatory system, concentrations of sessile phagocytes – 'reticular tissue' – may occur in addition to the circulating phagocytes; efficient clearance of circulating foreign particles from the blood is thus possible. For example, colloidal carbon particles and vertebrate erythrocytes rapidly accumulate in the sessile phagocytes of the dorsal diaphragm in locusts (A.M.L., unpublished observations).

Thus, for certain groups of invertebrates, we now know a considerable amount about the types of leucocyte and the roles they play in phagocytosis and encapsulation, but we do need to know more about these manifestations of cellular immunity in the less well-studied groups such as the coelenterates, echinoderms and tunicates. For example, we need more information about the naturally occurring responses as well as about the responses to artificial grafts. We need answers to questions such as: which leucocytes are first on the scene, functioning possibly as the recognition cells when a foreign object enters the body? Where several types of leucocyte occur within an individual, is there any evidence for co-operation between cell types? Are leucocytes involved in wound-healing as well as in the response to grafts? In addition, in order to help us understand immune recognition, we need to know which foreign objects stimulate a cellular response.

(b) *What do the cells recognize?*

Before we look at what invertebrate leucocytes recognize, in terms of either biotic or abiotic objects, we should first consider the ways in which recognition could occur. At its simplest level, recognition of non-self could be dependent on recognition of physical differences such as the overall charge or charge configuration of a surface, or of the degree of hydrophobicity of the surface. In the latter case, Van Oss & Gillman (1972), working with mammalian phagocytes, have shown that the greater the hydrophobicity of the surface of a bacterium relative to the surface of a phagocyte, the more easily will phagocytosis occur. The degree of hydrophobicity will, of course, be dependent on the number and arrangement of hydrophilic moieties such as carbohydrate chains, the terminal residues of which may also contribute to the surface charge. Recognition of physical differences in this way could thus give rise to a non-specific but effective recognition mechanism. Information on this type of response is essential in order to have a more precise base-line against which to compare our studies on specific immunity. A considerable amount of such information is already available for the vertebrate non-specific cellular response, so let us consider briefly what is known for invertebrates.

The types of abiotic objects which are phagocytosed or encapsulated (e.g. colloidal carbon, nylon thread, artificial 'sponge') have only been studied in detail in insects (see Table 5, Salt, 1970) and molluscs (Tripp, 1963; Sminia *et al.* 1974) and, even so, few people have sought to determine if there are abiotic objects which are not encapsulated. I am aware of only one paper which attempts to examine this problem, although in a slightly different context. Vinson (1974), investigating haemocytic encapsulation of Sephadex beads in the moth *Heliothis virescens* found that weakly acidic (C-50) beads were only thinly encapsulated compared with neutral (N-50) and weakly basic (A-50) beads. Further investigations on the effect of differing surface charge and of differing hydrophobicity of abiotic objects might well provide valuable information about the adhesion of invertebrate leucocytes to foreign objects. Differing rates of phagocytic uptake *in vitro* (reviewed by Anderson, 1975) or of phagocytic clearance *in vivo* (Pauley, Krassner & Chapman, 1971; Tyson, McKay & Jenkin, 1974) of different species of bacteria might suggest a certain degree of specificity in the cellular response. However, as noted above, such 'specificity' could arise from differences in physical characteristics alone. Further investigation of these differential phagocytic responses by invertebrate leucocytes might well provide results of particular interest for purposes of comparison with the vertebrate system.

In order to increase the acuity of recognition, it becomes necessary for the immune system to incorporate receptors which respond to specific 'non-self' molecular configurations. Lack of response to 'self' determinants cannot provide specificity although it could provide a graded response to objects with graded differences from the 'self' base-line. Of course, a tremendous amount is now known about the specificity of the vertebrate cellular response, but what do we know about the degree of specificity shown by invertebrate immune systems?

Some of the *in vitro* techniques borrowed from mammalian immunology, such as the investigation of the specificity of adhesion and rosette formation of vertebrate erythrocytes around invertebrate leucocytes (Cooper, 1973; Scott, 1971*a*; Ratcliffe & Gagen, 1976), or the study of cytotoxic reactions between the leucocytes of one individual and the 'target cells' of another have given us some information. Bertheussen (1979) has used the latter technique with sea-urchin cells to show that allogeneic recognition, followed by a cytotoxic response, occurs between leucocytes from two individuals; reciprocal 'killing' results, as shown by the release of chromium label into solution.

One other important method for examining cellular immune specificity *in vivo* has been to investigate the response to grafts or implants. On reading the literature, one finds that this exceptionally interesting field has so far been inadequately researched. In addition, so many of the results are open to other interpretations and, now that the concept of acute and chronic rejection of grafts by lower vertebrates seems to have been accepted, this has led to rather loose interpretation of results so that when a very delayed cellular response occurs, this may be used as evidence for immune recognition and rejection. A late cellular response may also

Table 2. *Recognition of allografts and xenografts*

Allografts = grafts between individuals of the same species.
Xenografts = grafts between individuals of different species.

	Allografts	Xenografts	References
Sponges	+	+	Curtis (1978), Moscona (1968)
Coelenterates	+	+	Hildemann *et al.* (1977), Theodor (1970)
Nemertines	−	±	Langlet & Bierne (1977)
Insects	−	±	Lackie (1979)
Crustacea	−	+	Crompton (1967)
Molluscs	−	+	Tripp (1963), Sminia *et al.* (1974)
Annelids	+	+	Cooper (1968), Dales (1978*a*)
Sipunculids	±	+	Cushing & Boraker (1975), Boiledieu & Valembois (1977)
Echinoderms	+	+	Hildemann & Dix (1972)
Tunicates	+	+	Freeman (1970), Tanaka (1975)

be stimulated by physiological incompatibility leading to death of the parasite (Yoeli, 1973), implant or graft, i.e. the cells would be recognizing dead, damaged or misaligned tissue (Parry, 1978) rather than recognizing differences in histo-compatibility. The type of implant used is important; some workers have im-planted pieces of tentacle or gill (minced tentacle, in one case) into the body cavity of recipients and have found that both auto- and allografts are recognized; presentation of tissue in this way must be akin to scarifying a skin graft or in-verting it in its graft bed! Finally in several phyla, leucocytes may also be involved in wound-healing, which complicates the analysis of the response to ectopic grafts. So, having borne in mind these important caveats, we can summarize the in-formation on recognition of grafts or implants as follows (Table 2).

Allogeneic recognition is found in the most primitive Metazoa, the Porifera and the Coelenterata, and in the Annelida and the Deuterostomes. As in the recogni-tion of histocompatibility by vertebrates, the extent of allogeneic recognition depends on the genetic inter-relationships of the individuals – for example, 'inbred' sea-urchins of the genus *Lytechinus* may not reject reciprocal allografts or may take much longer to do so than wild-type individuals (Coffaro & Hine-gardner, 1977). Allogeneic recognition by sipunculids has been tested *in vitro* by examining the cytotoxic reactions of leucocytes of one individual against 'erythrocytes' of another; allogeneic cytotoxicity was absent or considerably reduced depending on the geographic origin – and thus presumably the 'strain' – of the individual worms (Boiledieu & Valembois, 1977). For randomly trans-ferred allografts between neighbouring individuals of the sponge *Hymeniacidon perleve*, the frequency of acceptance is about 1 in 6; in colonial tunicates the occurrence of allogeneic acceptance may vary from 1 to 16% (Tanaka, 1975).

It now seems evident that allogeneic recognition does not occur in 3 Protosto-mate phyla, the Nemertinea, Arthropoda and Mollusca. Allografts, even between individuals from widely separated localities, are not recognized as foreign. A wide range of xenograft combinations has been tested only in insects and nemertines, and here the degree of recognition depends on the inter-specific combination and, to some extent, the phylogenetic inter-relationships of the species.

That 2 primitive groups, the Porifera and the Coelenterata, should be capable of allogeneic recognition, whereas many of the higher protostomates are not, might seem surprising. However, consideration of the life-style and form of these invertebrates provides a possible explanation. They have several points in common – they are sedentary as adults, they feed by filtering small particles from the water and consequently tend to grow in such a way as to present a larger surface area to the water flow, and they often live in crowded conditions surrounded by individuals of their own or other species. As a result of growth and crowding, there is grave danger of over-growth or of contact and fusion, and thus the loss of integrity of the individual sponge or colony. Selection will consequently act in favour of those individuals which recognize and react against contact by allogeneic and xenogeneic neighbours.

At present, we have no information on the nature of the effector cells involved in graft rejection in sponges and coelenterates. Although amoebocytes are found in coelenterate tissues, their presence has not so far been detected in coral grafts that are being rejected (Hildemann, Raison, Cheung, Hull, Akaka & Okamoto, 1977). Sponge archaeocytes (Table 1) accumulate in the fusion zones of grafts, but whether or not they interact with the foreign tissue is as yet unknown; it is not inconceivable that all sponge cells are capable of allogeneic recognition and production of inhibitory or cytotoxic factors.

(2) *Humoral immunity*

The presence of lysozyme in the haemolymph of insects and molluscs has been noted repeatedly (McHenery, Birkbeck & Allen, 1979). It is released from haemocytes during phagocytosis or encapsulation in molluscs (Cheng, Chorney & Yoshino, 1977; Kassim & Richards, 1978) and its release in insects can be stimulated by injection of endotoxin, bacteria and saline (Chadwick, 1975) although the magnitude of the response may depend on the nature of the material injected (Anderson & Cook, 1979). Other types of bactericidin have been found in various invertebrates; for example, in the blood of insects (Boman, Faye, Pye & Rasmuson, 1978; Natori, 1977), crustacea (Evans, Weinheimer, Painter, Acton & Evans, 1969), molluscs and tunicates (Johnson & Chapman, 1970*b*) and echinoderms (Wardlaw & Unkles, 1978).

The humoral factors upon which most attention is focused are the 'haemagglutinins', so-called because of their ability to agglutinate vertebrate erythrocytes *in vitro*. In addition, the agglutinins may agglutinate bacteria *in vitro* (Hardy, Fletcher & Olafsen, 1977; Pauley *et al.* 1971; Pistole, 1978; Scott, 1971*b*), the specificity and the titre of the agglutinins varying between the invertebrate species examined. It is now recognized that most of these agglutinins are lectin-like, i.e. molecules which have receptors for certain specific carbohydrate determinants (Sharon & Lis, 1972); accordingly, a considerable amount of work is in progress to determine their carbohydrate specificities.

Many invertebrates have been found to contain agglutinins and it seems to be possible to distinguish 2 types, those in the blood or coelomic fluid and those associated with the reproductive system. The lectins extracted from the albumen glands of certain snails react specifically with α-glycosidic (1–6)-bound galactose residues and N-acetylated amino sugars (Ishayama & Uhlenbruck, 1972; Renwrantz & Berliner, 1978). Agglutinins which react with N-acetyl neuraminic acid have been found in the blood of *Limulus* (Cohen, Rozenberg & Massaro, 1974), the oyster *Crassostrea gigas* (Hardy *et al.* 1977) and the lobster (Hall & Rowlands, 1974). The blood of other invertebrates may possess agglutinins with a wider range of specificities (McDade & Tripp, 1967; A.M.L., unpublished observations) as shown by agglutination inhibition tests. Agglutination of erythrocytes is not a totally reliable indicator of the presence or absence of an agglutinin, however, as binding without agglutination may occur, and specificities can only be confirmed effectively using the purified agglutinins. What then is the function of these agglutinins? Lectins were first isolated from plants and have now been found in vertebrates and invertebrates (Sharon & Lis, 1972); in none of these groups has the prime function of lectins been determined but many suggestions have been made. For example, they protect seeds against fungal attack; they are involved in transport of sugars, or of calcium for skeletal growth in molluscs (Tripp, 1974). Membrane-bound lectins could be important in morphogenetic relationships between cells; they may determine the inter-relationship between symbionts and tissues in the giant clam *Tridacna* (Baldo & Uhlenbruck, 1975). It is not inconceivable that such molecules containing a binding site with a certain specificity could become involved, perhaps initially by chance, with the organism's system of defence and come to be incorporated as 'recognition molecules' in the immune system. However, although we know that invertebrates have blood-borne, lectin-like molecules which may react with bacteria, vertebrate erythrocytes and some species of parasite *in vitro*, we still do not know whether this is their role *in vivo*. As Hardy *et al.* (1977) point out, with reference to the lectin from the haemolymph of *C. gigas*, 'the specificity directed towards sialic acids is inappropriate in that few bacteria and no fungi possess this sugar'. However, these authors suggest that other sterically related sugars on bacterial or fungal cell walls may also be able to bind to the lectin, if the sugars are present in sufficient concentration.

It might be expected that invertebrates which have lectins of similar specificity would exhibit rather similar reactions with respect to immune recognition and, so far, we cannot account for the observed differences. According to Parish's hypothesis (1977), self/non-self discrimination in invertebrates could be based on recognition of carbohydrate determinants by soluble or cell-bound oligomers of glycosyl-transferases. Although there is no evidence for secretion of glycosyl transferases into solution, the important point is made that enzymes of several specificities should be available, thus extending the range of recognition possible.

This field of research, the quest for humoral recognition factors, is thus both exciting and puzzling. However, we are still a long way from establishing, finally and unequivocally, whether or not immunological recognition is mediated via the blood-borne lectins.

(3) *Interactions between humoral and cellular components*

In order for humoral recognition factors to be immunologically effective, they must fulfil one or more of the following requirements: (*a*) inactivate bacteria rapidly by agglutination or lysis, (*b*) act as opsonins, forming a molecular link between a leucocyte and the foreign object, or (*c*) attach to the leucocyte to act as a cell-bound recognition molecule. Either of the last 2 mechanisms would enhance leucocyte adhesion and hence phagocytosis or encapsulation. There is certainly evidence for the opsonic effect of haemolymph on the uptake, both *in vitro* and *in vivo*, of bacteria and erythrocytes by phagocytes (Prowse & Tait, 1969; Renwrantz & Mohr, 1978; Tyson *et al.* 1974). Results of experiments by Tyson *et al.* (1974) suggest that bacteria are recognized by factors present both in the haemolymph and bound to the haemocyte membrane. A direct interaction between haemocytes and agglutinins has been shown by Amirante & Mazzalai (1978); the haemolymph of the cockroach *Leucophaea maderae* contains agglutinins for rabbit erythrocytes and, by means of fluorescent-labelled antibody prepared against the agglutinins, these authors have shown that they are present on the plasma membrane and in the cytoplasm of granulocytes and spherule cells. They suggest that the agglutinins, in addition to being membrane-bound, are released into the haemolymph.

Direct evidence for the opsonic effect of purified agglutinin on the phagocytosis of bacteria has been provided by Hardy *et al.* (1977) who also found, contrary to other workers, that exposure of oysters to bacteria stimulated an increased titre of agglutinin. Opsonic enhancement of phagocytosis by a lectin might require that the molecules bind not only to the surface of the phagocyte but also to the foreign particle. That lectins can act as molecular links between haemocytes and vertebrate erythrocytes has been shown in two interesting papers by Renwrantz & Cheng (1977*a,b*). They showed that human erythrocytes would form rosettes around haemocytes, from the snail *Helix pomatia*, which had been treated with various non-native agglutinins such as wheat germ agglutinin, concanavalin A and lectins from eels and from *Limulus*. In this system, since rosette formation was not followed by phagocytosis, a true opsonic effect was not demonstrated, but the experiments show that lectin-binding may enhance adhesion between leucocytes and certain biotic particles.

There is thus some support for the hypothesis that humoral factors, possibly the lectins, interact with leucocytes during recognition and adhesion to bacteria and erythrocytes. However, it should be remembered that there is also evidence against such interactions (Scott, 1971*a*; Smith & Ratcliffe, 1978) and this anomaly is unlikely to be explained until we know much more about the specificities of the different systems. In addition, we know virtually nothing about the interaction of humoral and cellular systems during immunological responses to grafts and parasites, and a comparative response may prove particularly useful here. For example, the oncosphere larvae of the cestode *Hymenolepis dimunuta* develop normally in the haemocoel of the locust *Schistocerca gregaria* but are recognized and encapsulated by haemocytes of the cockroach *Periplaneta americana* (Lackie, 1976). In addition, the oncospheres are agglutinated *in vitro* by diluted haemolymph from *P. americana* but not from *S. gregaria* (A.M.L., unpublished observations). Whether these humoral factors interact with the cellular response *in vivo* is, as yet, unknown.

II. DOES THE INVERTEBRATE IMMUNE RESPONSE SHOW INDUCTION AND MEMORY?

The vertebrate immune system can be adapted to respond to almost any foreign macromolecular substance. Once the response has been acquired its memory resides in the immune system of the animal, ready to be triggered into enhanced activity on subsequent contact with the specific antigen. In general, invertebrates appear unable to acquire a specific immune response – they will either respond at once or not at all; repeated infection with a specific foreign organism will not eventually stimulate immunity if none existed initially. There are some apparent exceptions, in which 'vaccination' with bacteria or endotoxin will stimulate protein synthesis (Boman *et al.* 1978), associated with induced production of bacteriocidin (reviewed by Chadwick, 1975). In most cases, further enhancement of the response was not shown after a second injection of the antigen (Boman *et al.* 1978), although Evans *et al.* (1969) found that the secondary and tertiary responses were increased in the lobster. It is possible that this inductive response is rather different from the acquired response of vertebrates – increased phagocytic activity, and therefore presumably an increased rate of production and synthesis of lysosomal enzymes can be stimulated by injection of carbon particles, indicating that the response is dormant rather than induced.

Immunological memory and an enhanced response against second-set grafts has been reported for sponges, corals (Hildemann *et al.* 1977), nemertines (Langlet & Bierne, 1977), annelids (Cooper, 1968) and echinoderms (Karp & Hildemann, 1976). However, in some cases the evidence is rather equivocal.

From the results of work from three different laboratories (Duprat,1964; Valembois, 1963; Cooper, 1968), it appears that secondary rejection of allo- or xenografts by earthworms is specific and enhanced. Specific memory must, therefore, be involved and can apparently be transferred by 'sensitized' coelomocytes (Hostetter & Cooper, 1974). However, Parry (1978) and Dales (1978*a*) have repeated the grafting experiments and neither of these authors finds evidence for accelerated rejection of second-set grafts. The problem lies in the interpretation of the results – Dales (1978*b*) discusses and re-evaluates Cooper's data and still comes to the same conclusion, that there is no evidence for memory and enhancement of the secondary response.

Echinoderms, because of their phylogenetic position, might be expected to provide us with useful information on specificity and memory in immunity, but the technical difficulties of grafting are obviously great. Although accelerated rejection of second-set grafts was claimed by Hildemann & Dix (1972), these grafts were placed on the recipient before the first-set graft was fully rejected, and therefore presumably while a large population of active coelomocytes was already present. Later experiments, with a small time interval between primary and secondary grafts, gave more convincing evidence of a memory component (Karp & Hildemann, 1976). Experiments by Coffaro & Hinegardner (1977) showed that rejection of primary allografts in the sea-urchin *Lytechinus pictus* occurred very rapidly (30–35 days). They used large numbers of urchins of known pedigree and, more important, their 'skin grafts' included the test as well as the overlying tissue. They found that calcium carbonate was usually deposited around the grafted test which then adhered firmly. The morphology of the stereom, the skeletal elements of the echinoderm test, is such that the soft tissues ramify throughout the stereom itself. Its inclusion in the graft could thus be particularly important. When allografts were rejected, the graft epidermis receded from the borders of the graft so that rejection was easily recognizable. Second-set grafts were given 2 months after rejection of the previous graft and were all rejected within 12 days. There is thus good evidence for some sort of memory but the response appears not to be specific as 'third-party' grafts from a third individual, given in place of the second-set graft, were also rejected in 12 days.

This particular field of invertebrate immunology, the search for memory and a specific, enhanced secondary response, is still really in its infancy although text-books have already been written on the subject (Cooper, 1976). There are technical difficulties and problems of interpretation. Choice of a suitable time-interval between the primary and secondary antigenic stimulus is of prime importance – too short an interval and the primary response will not have decreased to its resting level, so that a large population of sensitized cells is still available and ready to react. In the vertebrate immune response we would expect the secondary response to be accompanied by rapid division of the effector cells, so that increased numbers of reactive cells are available after stimulation. Convincing evidence for increased mitoses comes only from work on *Biomphalaria glabrata* (Lie, Heyneman & Jeong, 1976a) and on earthworms (Parry, 1976).

A good model system for studies on invertebrate immunity is the investigation of immunity to parasites. In general, if a healthy invertebrate host does not show an obvious immune response to a healthy parasite, it is unlikely that it will be immunized by subsequent exposures to the same species of parasite. Superimposed infections by one species are well-documented, and this suggests that the host has only 2 options – to react or not to react (see Section III). However, investigation of those host–parasite systems where the host normally responds by encapsulating the parasite, for example, should prove very valuable since the problems associated with the interpretation of the response to ectopic grafts are eliminated. In addition, parasites frequently have their own means of entry into the host, or are fairly easily injected.

At present there appears to be only one known example of a resistant host mounting an enhanced, fairly specific secondary response to a parasite. This system, which involves juveniles of the albino strain of the snail *B. glabrata* and larvae of the digenean *Echinostoma lindoense*, has been studied in detail by Lie, Heyneman & Lim (1975). If naturally resistant snails are exposed to normal miracidia or if susceptible snails are exposed to irradiated miracidia, the parasites penetrate the snail and migrate to the heart, but are then surrounded by clumps of amoebocytes. The sporocysts usually die as a result. If these snails are subsequently exposed to normal miracidia, the parasites are quickly destroyed by amoebocytes near the penetration area in the head-foot. This response is relatively specific – a primary infection with *E. lindoense* does not immunize the snail against challenge with *S. mansoni*, although a certain degree of protection is stimulated against *E. liei*. Interestingly, the amoebocyte-producing organ becomes enlarged and its cells show mitoses during this primary sensitization (Lie *et al.* 1976a).

Other attempts to immunize normally susceptible snails by infection with irradiated miracidia have been unsuccessful. *Lymnaea catascopium* infected with irradiated miracidia of *Schistosoma douthitti* and challenged with normal larvae, are just as likely to develop patent infections as are snails exposed only to normal miracidia (Loker, 1978), and Lie *et al.* (1975) were unable to find any significant protection against challenge with normal *S. mansoni* in *B. glabrata* previously infected with irradiated miracidia of *S. mansoni*. Consideration of the interactions of normal echinostome larvae with the immune system in naturally susceptible hosts (see Section III) suggests that the echinostome–snail relationship might be rather unusual. Investigation of these host–parasite systems in relation to immunity promises to be particularly fruitful and intriguing.

III. EVASION OF THE IMMUNE RESPONSE BY PARASITES

Now that we are beginning to understand more about the effectiveness of the immune mechanisms of invertebrates, we should be starting to ask questions such as: why do oysters succumb to bacterial infection? How can mosquitoes act as hosts for filarial worms? Why isn't *Fasciola hepatica* encapsulated in the snail? In other words, how do parasites and pathogens evade the immune response of their hosts?

The genetics of susceptibility and non-susceptibility of hosts will not be considered here since, although the genetics of several host–parasite systems have been quite well investigated (see review by Wakelin, 1978), it is often not possible to decide whether a change in susceptibility is due to immunological or physiological incompatibility. A plan is outlined below (Fig. 1) to show how the host's susceptibility is related to the inter-relationship between the parasite and the immune response. The scheme refers only to those parasites which have entered the body since there is no information available on whether parasites of the gut are affected by immunity.

Evasion of the host's immune system by a foreign organism can have only 2 possible results – death of the host due to damage by the invader, or a coexistence in a state of 'armed neutrality' between parasite and host. There are, of course, certain parasites, some mermithid nematodes and the larvae of parasitic wasps, which come into an intermediate category in that, although little mutual damage occurs in the early stages of their coexistence, the later stages of the relationship

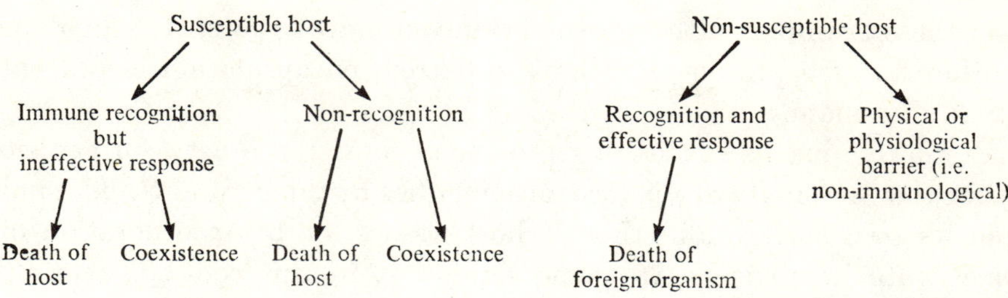

Fig. 1

destroy the host, either by rapacious devouring of its tissues or by a spectacular rupture of its surface.

The category of especial interest to parasitologists must be that of coexistence between parasite and host, either due to non-recognition or to recognition but interruption of the effector arm of the immune response. In both cases, examples come from the study of metazoan parasites in insects and in the mollusc, *B. glabrata*.

Evidence for survival of the parasite by interruption of the effector arm of the immune response comes from work on the insect parasitoid *Pseudeucoila bochei*, which lays its eggs in larvae of the dipteran *Drosophila* spp. Walker (quoted in Nappi, 1975) found that larvae of a strain of *P. bochei* suppressed haemocyte transformation and thus capsule formation in *D. melanogaster*. This was followed up by Nappi and his co-workers (see review by Nappi, 1975) who showed that transformed haemocytes were found in parasitized *D. algonquin*, but the cells were apparently incapable of reacting with and encapsulating the parasite. The protective factor produced by this strain of *P. bochei* could also protect a different parasite, *P. mellipes*, against encapsulation when both parasites were present in the same host.

A similar effect has been found in the mollusc *B. glabrata* parasitized by *E. lindoense* (Lie & Heyneman, 1976; Lie, Heyneman & Jeong. 1976*b*). As mentioned previously (Section II), sporocysts of *E. lindoense* are encapsulated by amoebocytes in the heart of resistant *B. glabrata*. However, sporocysts occasionally escape from the capsule and continue their development in the ovotestis of the snail. When this happens the snail loses its immunity to a challenge infection and becomes more susceptible to further infection. A challenge infection stimulates increased activity in the amoebocyte-producing organ but the amoebocytes are no longer attracted to the invading parasite and accumulate in the connective tissue of the digestive gland. Lie & Heyneman (1976) suggest that 'continued survival of the escaped parasite inhibits or interferes with the ability of the amoebocytes to recognize them or be attracted to them'. Their subsequent hypothesis, that *living* sporocysts inhibit or interfere with the snail's immune response, thus providing protection for the parasites of both primary and secondary infections, was supported by the results of heterologous infections of *S. mansoni* and *Paryphostomum segregatum* with *E. lindoense* in which living sporocysts of either species partially protected irradiated miracidia of *E. lindoense* (which are normally encapsulated) against encapsulation, both in naturally resistant snails and in snails with acquired resistance (Lie & Heyneman, 1977; Lie *et al.* 1976*b*). Thus, in these examples, although the haemocytes are apparently stimulated and recognition may be presumed to have occurred, encapsulation is prevented in some unknown manner.

An alternative means of ensuring prolonged coexistence between parasite and host is for the parasite to evade recognition either by inherent antigenic similarity between its own surface and that of host tissues, or by incorporation of host molecules onto its surface. Antigenic similarity, in this context, refers to the discrimination of differences in histocompatibility by haemocytes, and not to the much more sensitive discrimination shown by the vertebrate immune system. There is now evidence to suggest that when the envelope around larvae of the acanthocephalan *Moniliformis dubius* is fully formed, it possesses some antigenic similarity to the tissues of its normal host, the cockroach *Periplaneta americana* (J. Lackie, 1975). Such similarity does not derive from incorporation of host molecules, for the envelope is largely parasite-derived (Lackie & Lackie, 1979). Larvae of the cestode *H. diminuta* develop unencapsulated in several insect species and apparently share antigenic similarity with their hosts' tissues (A. Lackie, 1976).

This postulated existence of antigenic similarity need not seem such a revolutionary concept if one considers that the invertebrate immune system is based on the recognition of rather gross differences of physical and/or chemical parameters (compared with the exquisite sensitivity of vertebrate immune recognition). There will then be a certain range of histocompatibility within which the surfaces are recognized as similar. Thus, the possibility will always exist that the tissue surfaces of 2 organisms will have such a configuration as to appear immunologically compatible. That such compatibilities do exist is indicated by non-recognition of allografts and of certain xenografts (see Section I, 3), and there is no reason to suppose that such compatibility could not also be expressed by the surface of a parasite.

In relation to this, work on the serum glycosubstances of *B. glabrata* and the surface carbohydrates of *S. mansoni* larvae may be particularly relevant. Yoshino, Cheng & Renwrantz (1977) have shown that the surface of the miracidium has carbohydrate determinants which react specifically with 3 different lectins and with antibody prepared against human Group A erythrocytes. In the serum of *B. glabrata* there are sugar-containing macromolecules, some of them glycoproteins, which react with these same 4 reagents (Stanislawski, Renwrantz & Becker, 1976). Thus, the surface of the miracidium and the serum of its host possess molecules with similar carbohydrate determinants. In addition, antibody prepared against whole haemolymph also reacts with the surface of miracidia (Yoshino & Cheng, 1978). In contrast, Stein & Basch (1979), drawing parallels with the evasion mechanisms of adult *S. mansoni* in mammals, postulate that evasion of cellular recognition in the snail occurs, not by inherent antigenic similarity but rather by bound agglutinin masking the larvae from detection. These authors adsorbed haemolymph with Sephadex G-150 gel and the fraction thus isolated, which they suggest was purified agglutinin, was used to raise antibody. The fluorescence-labelled antibody did not react with the surfaces of free miracidia or of cultured sporocysts, but did react with the surfaces of sporocysts and cercariae dissected from the snail. Why adsorption of agglutinin should prevent rather than enhance attack by the host's haemocytes has yet to be explained and it is perhaps unprofitable to speculate further until the relative binding affinities of agglutinin for 'self' and 'non-self' have been determined.

A final question remains. Why are arthropods and molluscs such popular hosts for parasites? Apart from the fact that these hosts are numerous and ubiquitous, it is interesting to speculate on the inter-relationship between the parasite and the host's immune response. The ability of the immune recognition systems of molluscs and arthropods to discriminate difference appears to be much lower than that of annelids (Section I, 3), allogeneic recognition is absent and in some cases xenogeneic recognition is reduced. One can thus easily imagine a situation in which a parasite can slip past the recognition mechanism. Obviously, for the parasite merely to gain entry and remain unmolested by the immune response is not sufficient for a long-term, well-balanced host–parasite system, and selection pressures will subsequently favour mutual adaptations leading towards a state of balanced coexistence.

REFERENCES

AMIRANTE, G. A. & MAZZALAI, F. G. (1978). Synthesis and localisation of hemoagglutinins in hemocytes of cockroach *Leucophaea maderae* L. *Developmental and Comparative Immunology* **2**, 435–40.

ANDERSON, R. S. (1971). Cellular responses to foreign bodies in the tunicate *Molgula manhattensis* (DeKay). *Biological Bulletin, Woods Hole* **141**, 91–8.

ANDERSON, R. S. (1975). Phagocytosis by invertebrate cells *in vitro*: biochemical events and other characteristics compared with vertebrate phagocytic systems. In *Invertebrate Immunity* (ed. K. Maramorosch and R. E. Shope). New York and London: Academic Press.

ANDERSON, R. S. & COOK, M. L. (1979). Induction of lysozyme-like activity in the hemolymph and hemocytes of an insect, *Spodoptera eridania*. *Journal of Invertebrate Pathology* **33**, 197–203.

ASTON, W. P. & CHADWICK, J. S. (1978). Time and dose studies of the effect of cobra venom factor on the *in vivo* immune response in *Galleria mellonella* to *Pseudomonas aeruginosa*. *Developmental and Comparative Immunology* **2**, 425–34.

BALDO, B. A. & UHLENBRUCK, G. (1975). Tridacnin, a potent anti-galactan precipitin from the hemolymph of *Tridacna maxima* (Röding). In *Immunologic Phylogeny* (ed. W. H. Hildemann and A. A. Benedict). Advances in Experimental Medicine and Biology, vol 64. New York and London: Plenum Press.

BERTHEUSSEN, K. (1979). The cytotoxic reaction in allogeneic mixtures of echinoid phagocytes. *Experimental Cell Research* **120**, 373–82.

BERTHEUSSEN, K. & SELJELID, R. (1978). Echinoid phagocytes *in vitro*. *Experimental Cell Research* **111**, 401–12.

BOILEDIEU, D. & VALEMBOIS, P. (1977). Natural cytotoxic activity of sipunculid leukocytes on allogenic and xenogenic erythrocytes. *Developmental and Comparative Immunology* **1**, 207–16.

BOMAN, H. G., FAYE, I., PYE, A. & RASMUSON, T. (1978). The inducible immunity system of Giant Silk Moths. In *Comparative Pathobiology*, vol. 4 (ed. L. A. Bulla and T. C. Cheng). New York and London: Plenum Press.

CHADWICK, J. S. (1975). Hemolymph changes with infection or induced immunity in insects and ticks. In *Invertebrate Immunity* (ed. K. Maramorosch and R. E. Shope). New York and London: Academic Press.

CHENG, T. C. (1975). Functional morphology and biochemistry of molluscan phagocytes. *Annals of the New York Academy of Sciences* **266**, 343–79.

CHENG, T. C. (1976). Beta-glucuronidase in the serum and hemolymph cells of *Mercenaria mercenaria* and *Crassostrea virginica* (Mollusca: Pelecypoda). *Journal of Invertebrate Pathology* **27**, 125–8.

CHENG, T. C., CHORNEY, M. J. & YOSHINO, T. P. (1977). Lysozyme-like activity in the hemolymph of *Biomphalaria glabrata* challenged with bacteria. *Journal of Invertebrate Pathology* **29**, 170–4.

CHENG, T. C. & GARRABRANT, T. A. (1977). Acid phosphatase in granulocytic capsules formed in strains of *Biomphalaria glabrata* totally and partially resistant to *Schistosoma mansoni*. *International Journal for Parasitology* **7**, 467–72.

COFFARO, K. A. & HINEGARDNER, R. T. (1977). Immune response in the sea urchin *Lytechinus pictus*. *Science* **197**, 1389–90.

COHEN, E., ROZENBERG, M. & MASSARO, E. J. (1974). Agglutinins of *Limulus polyphemus* (Horseshoe crab) and *Birgus latro* (coconut crab). *Annals of the New York Academy of Sciences* **234**, 28–33.

COOPER, E. L. (1968). Transplantation immunity in annelids. I. Rejection of xenografts exchanged between *Lumbricus terrestris* and *Eisenia foetida*. *Transplantation* **6**, 322–37.

COOPER, E. L. (1973). Evolution of cellular immunity. In '*Non-specific*' *Factors Influencing Host Resistance* (ed. W. Braun and J. Ungar). Basel: Karger.

COOPER, E. L. (1976). *Comparative Immunology*. New Jersey: Prentice-Hall.

CROMPTON, D. W. T. (1967). Studies on the haemocytic reaction of *Gammarus* species and its relationship to *Polymorphus minutus* (Acanthocephala). *Parasitology* **57**, 389–401.

CURTIS, A. S. G. (1978). Individuality and graft rejection in sponges or, a cellular basis for individuality in sponges. In *Biology and Systematics of Colonial Organisms* (ed. G. Larwood and B. R. Rosen). London and New York: Academic Press.

CUSHING, J. E. & BORAKER, D. K. (1975). Some specific aspects of cell-surface recognition by sipunculid coelomocytes. In *Immunologic Phylogeny*. Advances in Experimental Medicine and Biology, vol. 64 (ed. W. H. Hildemann and A. A. Benedict). New York and London: Plenum Press.

DALES, R. P. (1978a). The basis of graft rejection in the earthworms *Lumbricus terrestris* and *Eisenia foetida*. *Journal of Invertebrate Pathology* **32**, 264–77.

DALES, R. P. (1978b). Second-set graft rejections: do they occur in invertebrates? *Symposium of the Society for Experimental Biology* **32**, 203–19.

DAY, N. K. B., GEWURZ, H., JOHANNSEN, R., FINSTAD, J. & GOOD, R. A. (1970). Complement and complement-like activity in lower vertebrates and invertebrates. *Journal of Experimental Medicine* **132**, 941–50.

DUPRAT, P. (1964). Mise en evidence de reaction immunitaire dans les homogreffes de paroi du corps chez le lombricien *Eisenia foetida typica*. *Comptes rendus hebdomadaires des séances de l'Académie des Sciences, Paris* **259**, 4177–9.

EVANS, E. E., WEINHEIMER, P. F., PAINTER, B., ACTON, R. T. & EVANS, M. L. (1969). Secondary and tertiary responses of the induced bactericidin from the West Indian spiny lobster, *Panulirus argus*. *Journal of Bacteriology* **98**, 943–6.

FONTAINE, A. R. & LAMBERT, P. (1977). The fine structure of the leucocytes of the holothurian, *Cucumaria miniata*. *Canadian Journal of Zoology* **55**, 1530–44.

FREEMAN, G. (1970). Transplantation specificity in echinoderms and lower chordates. *Transplantation Proceedings* **2**, 236–9.

GRIMSTONE, W. B., ROTHERAM, S. & SALT, G. (1967). An electron microscope study of capsule formation by insect blood cells. *Journal of Cell Science* **2**, 281–92.

GUPTA, A. R. (1979). *Insect Hemocytes: Development, Forms, Functions and Techniques*. Cambridge University Press.

HALL, J. L. & ROWLANDS, D. T. (1974). Heterogeneity of lobster agglutinins. II. Specificity of agglutinin-erythrocyte binding. *Biochemistry* **13**, 828–33.

HARDY, S. W., FLETCHER, T. C. & OLAFSEN, J. A. (1977). Aspects of cellular and humoral defence mechanisms in the Pacific oyster, *Crassostrea gigas*. In *Developmental Immunobiology* (ed. J. B. Solomon and J. D. Horton). Amsterdam: Elsevier/North-Holland.

HARRIS, K. R. (1975). The fine structure of encapsulation in *Biomphalaria glabrata*. *Annals of the New York Academy of Science* **266**, 446–64.

HILDEMANN, W. H. & DIX, T .G. (1972). Transplantation reactions of tropical Australian echinoderms. *Transplantation* **14**, 624–33.

HILDEMANN, W. H., RAISON, R. L., CHEUNG, G., HULL, C. J., AKAKA, L. & OKAMOTO, J. (1977). Immunological specificity and memory in a scleractinian coral. *Nature, London* **270**, 219–23.

HOSTETTER, R. K. & COOPER, E. L. (1972). Coelomocytes as effector cells in earthworm immunity. *Immunological Communications* **1**, 155–83.

HOSTETTER, R. K. & COOPER, E. L. (1974). Earthworm coelomocyte immunity. In *Contemporary Topics in Immunobiology*, vol. 4 (ed. E. L. Cooper). Plenum Press.

ISHAYAMA, I. & UHLENBRUCK, G. (1972). Further studies on the specificity of the anti-A agglutinin from *Helix pomatia*. *Comparative Biochemistry and Physiology* A **42**, 269–76.

JOHNSON, P. T. & CHAPMAN, F. A. (1970a). Infection with diatoms and other microorganisms in sea urchin spines (*Stronglyocentrotus franciscanus*). *Journal of Invertebrate Pathology* **16**, 268–76.

JOHNSON, P. T. & CHAPMAN, F. A. (1970b). Comparative studies on the *in vitro* response of bacteria to invertebrate body fluids. II. *Aplysia californica* (sea-hare) and *Ciona intestinalis* (tunicate). *Journal of Invertebrate Pathology* **16**, 127–38.

KARP, R. D. & HILDEMANN, W. H. (1976). Specific allograft reactivity in the seastar *Dermasterias imbricata*. *Transplantation* **22**, 434–9.

KASSIM, O. O. & RICHARDS, C. S. (1978). *Biomphalaria glabrata*: lysozyme activities in the hemolymph, digestive gland and headfoot of the intermediate host of *Schistosoma mansoni*. *Experimental Parasitology* **46**, 218–24.

LACKIE, A. M. (1976). Evasion of the haemocytic defence reaction of certain insects by larvae of *Hymenolepis diminuta* (Cestoda). *Parasitology* **73**, 97–107.

LACKIE, A. M. (1979). Cellular recognition of foreign-ness in two insect species, the American cockroach and the desert locust. *Immunology* **36**, 909–14.

LACKIE, A. M. & LACKIE, J. M. (1979). Evasion of the insect immune response by *Moniliformis dubius* (Acanthocephala): further observations on the origin of the envelope. *Parasitology* **79**, 297–301.

LACKIE, J. M. (1975). The host specificity of *Moniliformis dubius*, (Acanthocephala), a parasite of cockroaches. *International Journal for Parasitology* **5**, 301–7.

LANGLET, C. & BIERNE, J. (1977). The immune response to xenografts in nemertines of the genus *Lineus*. In *Developmental Immunobiology* (ed. J. B. Solomon and J. D. Horton). Amsterdam and Oxford: Elsevier/North-Holland.

LIE, K. J. & HEYNEMAN, D. (1976). Studies on resistance in snails. 6. Escape of *Echinostoma lindoense* sporocysts from encapsulation in the snail heart and subsequent loss of the host's ability to resist infection by the same parasite. *Journal of Parasitology* **62**, 298–302.

LIE, K. J. & HEYNEMAN, D. (1977). *Schistosoma mansoni, Echinostoma lindoense*, and *Paryphostomum segregatum*: interference by trematode larvae with acquired resistance in snails, *Biomphalaria glabrata*. *Experimental Parasitology* **42**, 343–7.

LIE, K. J., HEYNEMAN, D. & JEONG, K. H. (1976a). Studies on resistance in snails. 4. Induction of ventricular capsules and changes in the amoebocyte-producing organ during sensitization in *Biomphalaria glabrata* snails. *Journal of Parasitology* **62**, 286–91.

LIE, K. J., HEYNEMAN, D. & JEONG, K. H. (1976b). Studies on resistance in snails. 7. Evidence of interference with the defence reaction in *Biomphalaria glabrata* by trematode larvae. *Journal of Parasitology* **62**, 608–15.

LIE, K. J., HEYNEMAN, D. & LIM, H. K. (1975). Studies on resistance in snails: specific resistance induced by irradiated miracidia of *Echinostoma lindoense* in *Biomphalaria glabrata* snails. *International Journal for Parasitology* **5**, 627–31.

LOKER, E. S. (1978). *Schistosomatium douthitti*: exposure of *Lymnaea catascopium* to irradiated miracidia. *Experimental Parasitology* **46**, 134–40.

McDADE, J. E. & TRIPP, M. R. (1967). Mechanisms of agglutination of red blood cells by oyster hemolymph. *Journal of Invertebrate Pathology* **9**, 523–30.

McHENERY, J. G., BIRKBECK, T. H. & ALLEN, J. A. (1979). Occurrence of lysozyme in marine bivalves. *Comparative Biochemistry and Physiology* B **63**, 25–8.

MOSCONA, A. A. (1968). Cell aggregation: properties of specific cell-ligands and their role in the formation of multicellular systems. *Developmental Biology* **18**, 250–77.

NAPPI, A. J. (1975). Parasite encapsulation in insects. In *Invertebrate Immunity* (ed. K. Maramorosch and R. E. Shope). New York and London: Academic Press.

NATORI, S. (1977). Bactericidal substance induced in the haemolymph of *Sarcophaga peregrina* larvae. *Journal of Insect Physiology* **23**, 1169–73.

PARISH, C. R. (1977). Simple model for self- non-self discrimination in invertebrates. *Nature, London* **267**, 711–13.

PARRY, M. J. (1976). Evidence of mitotic division of coelomocytes in the normal, wounded and grafted earthworm *Eisenia foetida*. *Experientia* **32**, 449–50.

PARRY, M. J. (1978). Survival of body wall autografts, allografts and xenografts in the earthworm, *Eisenia foetida*. *Journal of Invertebrate Pathology* **31**, 383–8.

PAULEY, G. B., KRASSNER, S. M. & CHAPMAN, F. A. (1971). Bacterial clearance in the California seahare *Aplysia californica*. *Journal of Invertebrate Pathology* **18**, 227–39.

PISTOLE, T. G. (1978). Broad-spectrum bacterial agglutinating activity in the serum of the Horseshoe crab, *Limulus polyphemus*. *Developmental and Comparative Immunology* **2**, 65–76.

POINAR, G. O. & LEUTENEGGER, R. (1971). Ultrastructural investigations of the melanization process in *Culex pipiens* (Culicidae) in response to a nematode. *Journal of Ultrastructural Research* **36**, 149–58.

PROWSE, R. H. & TAIT, N. N. (1969). *In vitro* phagocytosis by amoebocytes from the haemolymph of *Helix aspersa* (Müller). 1. Evidence for opsonic factor(s) in the serum. *Immunology* **17**, 437–43.

RATCLIFFE, N. A. & GAGEN, S. J. (1976). Cellular defense reactions of insect hemocytes *in vivo*: nodule formation and development in *Galleria mellonella* and *Pieris brassicae* larvae. *Journal of Invertebrate Pathology* **28**, 373–82.

REINISCH, C. L. & BANG, F. B. (1971). Cell recognition: reaction of the sea-star *Asterias vulgaris* to the injection of amoebocytes of sea urchin *Arbacia punctata*. *Cellular Immunology* **2**, 496–51.

RENWRANTZ, L. & BERLINER, U. (1978). A galactose-specific agglutinin, a blood-group H active polysaccharide and a trypsin inhibitor in albumen glands and eggs of *Arianta arbustorum* (Helicidae). *Journal of Invertebrate Pathology* **31**, 171–9.

RENWRANTZ, L. R. & CHENG, T. C. (1977a). Identification of agglutinin receptors on hemocytes of *Helix pomatia*. *Journal of Invertebrate Pathology* **29**, 88–96.

RENWRANTZ, L. R. & CHENG, T. C. (1977b). Agglutinin-mediated attachment of erythrocytes to hemocytes of *Helix pomatia*. *Journal of Invertebrate Pathology* **29**, 97–100.

RENWRANTZ, L. & MOHR, W. (1978). Opsonizing effect of serum and albumin gland extracts on the elimination of human erythrocytes from the circulation of *Helix pomatia*. *Journal of Invertebrate Pathology* **31**, 164–70.

ROWLEY, A. F. & RATCLIFFE, N. A. (1979). An ultrastructural and cytochemical study of the interaction between latex particles and the haemocytes of the wax moth *Galleria mellonella in vitro*. *Cell and Tissue Research* **199**, 127–37.

SALT, G. (1970). *Cellular Defence Reactions of Insects*. Cambridge University Press.

SCHMIT, A. R. & RATCLIFFE, N. A. (1978). The encapsulation of Araldite implants and the recognition of foreignness in *Clitumnus extradentatus*. *Journal of Insect Physiology* **24**, 511–22.

SCOTT, M. T. (1971a). Recognition of foreignness in invertebrates. II. *In vitro* studies of cockroach phagocytic haemocytes. *Immunology* **21**, 817–28.

SCOTT, M. T. (1971b). A naturally-occurring haemagglutinin in the haemolymph of the American cockroach. *Archives de zoologie experimentale et generale* **112**, 73–80.

SHARON, N. & LIS, H. (1972). Lectins: cell-agglutinating and sugar-specific proteins. *Science* **177**, 949–59.

SMINIA, T. (1972). Structure and function of blood and connective-tissue cells of the freshwater pulmonate *Lymnaea stagnalis* studied by electron microscopy and enzyme histochemistry. *Zeitschrift für Zellforschung* **130**, 497–526.

SMINIA, T., BORGHARDT-REINDERS, E. & VAN DE LINDE, A. W. (1974). Encapsulation of foreign materials experimentally introduced into the freshwater snail *Lymnaea stagnalis*: an electron microscopic and autoradiographic study. *Cell and Tissue Research* **153**, 307–26.

SMITH, M. J. (1970). The blood cells and tunic of the ascidian *Holocynthia aurantium*. I. Hematology, tunic morphology, and partition of cells between blood and tunic. *Biological Bulletin, Woods Hole* **138**, 354–78.

SMITH, V. J. & RATCLIFFE, N. A. (1978). Host defence reactions of the shore crab, *Carcinus maenas* (L.) *in vitro*. *Journal of the Marine Biological Association, UK* **58**, 367–79.

SOLANGI, M. A. & LIGHTNER, D. V. (1976). Cellular inflammatory response of *Penaeus aztecus* and *P. setiferus* to the pathogenic fungus, *Fusarum* sp; isolated from the California brown shrimp, *P. californiensis*. *Journal of Invertebrate Pathology* **27**, 77–86.

STANISLAWSKI, E., RENWRANTZ, L. & BECKER, W. (1976). Soluble blood group reactive substances in the hemolymph of *Biomphalaria glabrata* (Mollusca). *Journal of Invertebrate Pathology* **28**, 301–8.

STEIN, P. C. & BASCH, P. F. (1979). Purification and binding properties of hemagglutinin from *Biomphalaria glabrata*. *Journal of Invertebrate Pathology* **33**, 10–18.

TANAKA, K. (1975). Allogeneic distinction in *Botryllus primigenus* and in other colonial ascidians. In *Immunologic Phylogeny* (ed. W. H. Hildemann and A. A. Benedict). Advances in Experimental Medicine and Biology, vol. 64. New York and London: Plenum Press.

THEODOR, J. L. (1970). Distinction between 'self' and 'not-self' in lower invertebrates. *Nature, London* **227**, 690–2.

TRIPP, M. R. (1963). Cellular responses of mollusks. *Annals of the New York Academy of Sciences* **113**, 467–74.

TRIPP, M. R. (1974). Molluscan immunity. *Annals of the New York Academy of Sciences* **234**, 23–7.

TYSON, C. J., MCKAY, D. & JENKIN, C. R. (1974). Recognition of foreign-ness in the freshwater crayfish, *Parachaeraps bicarinatus*. In *Contemporary Topics in Immunobiology*, vol. 4, (ed. E. L. Cooper). Plenum Press.

VALEMBOIS, P. (1963). Recherches sur la nature de la reaction antigreffe chez le lombricien *Eisenia foetida* Savigny. *Comptes rendus hebdomadaires des seances de l'Academie des Sciences, Paris* **257**, 3489–90.

VAN OSS, C. J. & GILLMAN, C. F. (1972). Phagocytosis as a surface phenomenon. I. *Journal of the Reticuloendothelial Society* **12**, 283–92.

VINSON, S. B. (1974). The role of the foreign surface and female parasitoid secretions on the immune response of an insect. *Parasitology* **68**, 27–33.

WAKELIN, D. (1978). Genetic control of susceptibility and resistance to parasitic infection. *Advances in Parasitology*, vol. 16 (ed. W. H. R. Lumsden, R. Muller and J. R. Baker). New York and London: Academic Press.

WARDLAW, A. C. & UNKLES, S. E. (1978). Bactericidal activity of coelomic fluid from the sea urchin *Echinus esculentus*. *Journal of Invertebrate Pathology* **32**, 125–34.

YOELI, M. (1973). *Plasmodium berghei*: mechanisms and sites of resistance to sporogonic development in different mosquitoes. *Experimental Parasitology* **34**, 448–58.

YOSHINO, T. P. & CHENG, T. L. (1978). Snail host-like antigens associated with the surface membranes of *Schistosoma mansoni* miracidia. *Journal of Parasitology* **64**, 752–3.

YOSHINO, T. P., CHENG, T. C. & RENWRANTZ, L. R. (1977). Lectin and human blood group determinants of *Schistosoma mansoni*–alteration following *in vitro* transformation of miracidium to mother sporocyst. *Journal of Parasitology* **63**, 818–24.

Parasitology (1980), **80**, 571–579

Some implications of a sexual cycle in *Theileria*

A. D. IRVIN *and* C. D. H. BOARER

*International Laboratory for Research on Animal Diseases,
P.O. Box 30709, Nairobi, Kenya*

(*Accepted* 29 *November* 1979)

SUMMARY

Recent microscopic studies of theilerial parasites in ticks appear to provide convincing evidence for a sexual cycle for these parasites. This evidence would remove most of the objections to placing piroplasms in the same class as malarial and coccidial parasites. It is suggested, therefore, that the class Piroplasmasida (Piroplasmea) be abolished and that piroplasms be located in a sub-order Piroplasmorina alongside Haemosporina within the class Sporozoa (Teleospora).

If the close relationship of theilerial and malarial parasites is correct then by analogy meiotic or reduction division of *Theileria* probably occurs during sporogony in the tick salivary gland acinar cells. This means that all bovine stages are haploid. If these can be cloned and characterized by isoenzyme or drug resistance markers, then genetic recombinant studies, similar to those performed with malarial and coccidial parasites, could be conducted.

The potential of genetic recombination greatly increases the scope of the parasite for developing new variants. This potential is greatest in areas where there is a variety of tick and host species which apply different types of selection pressure to parasites. Atypical strains of *Theileria* are usually isolated from such areas, and although the characteristics of these strains may not be expressed in the wild, unnatural selection by laboratory passage may reveal their presence. The apparent lability of such strains may simply be cloning of different genetic recombinants.

If different strains can be cloned and characterized it may be possible by genetic recombination techniques to develop new clones which retain desired characters for use as avirulent protective vaccines, but from which undesirable characters have been eliminated.

INTRODUCTION

Recent work on the life-cycle of *Theileria* in ticks appears finally to have resolved the question of whether or not a sexual stage occurs during the development of the parasite. Schein and his colleagues have studied the development of both *T. annulata* and *T. parva* in *Hyalomma anatolicum excavatum* (Schein, 1975; Schein, Buscher & Friedhoff, 1975; Mehlhorn & Schein 1976, 1977) and *T. parva* in *Rhipicephalus appendiculatus* (Schein, Warnecke & Kirmse, 1977) and for both parasites describe the maturation of 'male' forms (microgamonts and microgametes), the presence of 'female' forms (macrogametes) and the maturation and development of 'zygotes'. They thus commit themselves firmly in favour of a sexual cycle and although they have so far been unable to demonstrate syngamy of 'male' and 'female' forms, their careful and detailed studies would seem to support the concept of sex in *Theileria* which was originally proposed by Koch (1906) and which since then has received an almost equal share of antagonists and protagonists.

On the assumption that the recent interpretations are correct there are a number of implications which could have significant bearing both on past findings and on future research. Some of these implications are discussed below.

TAXONOMY

The controversy surrounding the placing of *Theileria* and *Babesia* in the higher taxa has largely revolved around the question of a sexual cycle. To many of the earlier workers the similarities between the life-cycles of malarial parasites and piroplasms seemed to justify placing them closely together under Haemosporidia in the class Sporozoa. However, the uncertainty regarding a possible sexual cycle led Levine (1971) to state 'In the absence of evidence to the contrary, we can conclude that sex does not occur in the piroplasms'. This was a stand that had earlier been taken by the Committee on Taxonomy and Taxonomic Problems of the Society of Protozoologists (Honigberg, Balamuth, Bovee, Corliss, Gojdics, Hall, Kudo, Levine, Loeblich, Weiser & Wenrich, 1964), and as a consequence piroplasms were removed from the Sporozoa. However, not all protozoologists accepted this action and, in particular, the work of Riek (1964) on *Babesia bigemina* suggested that a sexual cycle did occur.

Since the definitive statements by Honigberg *et al.* (1964) the piroplasms have drifted uncertainly between various classes, sub-classes and appendages (Brocklesby, 1978) further from or nearer to malarial parasites, according to the whims and doubts of the pundits, until ultrastructural studies established firmly the links between piroplasms and malarial and coccidial parasites. As a result all 3 groups were placed into the sub-phylum Apicomplexa (Levine, 1971). This did not resolve the question of lower taxonomic placings but re-acknowledged that a close link existed between piroplasms and malaria. Two classes are described within the Apicomplexa, the Sporozoasida and the Piroplasmasida, the difference being based on the presence and absence respectively of a sexual cycle (Levine, 1973).

Baker (1977) adopted a similar, but more traditionalist approach in preferring the name Sporozoa for the sub-phylum Apicomplexa but, in common with Levine, proposed 2 classes: Teleospora and Piroplasmea, splitting them essentially on the controversy of the sexual cycle. We feel that both these proposals are incorrect since it now appears clear that a sexual cycle does occur in the piroplasms. We would propose, therefore, returning the piroplasms to what many protozoologists would intuitively regard as their rightful place alongside the malaria parasites; abolishing the class Piroplasmasida (Piroplasmea) and assigning piro-

Table 1. *Classification of theilerial parasites under the Levine or Baker system*

	Levine	Baker
Phylum	Protozoa	Protozoa
Sub-phylum	Apicomplexa	Sporozoa
Class	Sporozoasida	Teleospora
Sub-class	Coccidiasina	Coccidia
Order	Eucoccidiorida	Eucoccida
Sub-order	Piroplasmorina	Piroplasmorina
Family	Theileriidae	Theileriidae
Genus	*Theileria*	*Theileria*

plasms to a sub-order Piroplasmorina. Within this system theilerial parasites could then be classified under either the Levine or the Baker system as shown in Table 1.

In suggesting this classification we lean more towards the Baker system but are less concerned with the finer points of nomenclature and semantics than with rationalizing the relationship between malarial parasites and piroplasms now that it is apparent that taxonomic separation, based on a sexual cycle, is no longer tenable. However, we would concede that separation on the basis of gamete structure may be necessary, in which case we would propose re-instating the order Piroplasmida, but would baulk at the idea of separation at Class level. At this stage we believe that the structure of theilerial gametes is insufficiently known to justify this step and prefer to adopt the stand described.

GENETICS

Probably the most important implication of a sexual cycle in *Theileria* is the potentiality for genetic mixing and recombination. This immediately increases the scope of the parasite for the development of different strains in the face of selection pressures.

From the studies of Schein and his colleagues it seems clear that genetic recombination of theilerial gametes occurs by a process of anisogamy in the gut of the tick within 1–4 days of repletion on the bovine host. Although syngamy was not recorded, the presence of dimorphic 'male' and 'female' gametes, followed by the subsequent appearance of zygotes, suggests this interpretation is correct. If so, zygote formation presumably results in the production of a diploid parasite which undergoes maturation in the tick to a motile kinete. However, the mechanism whereby the parasite undergoes meiotic or reduction division to form haploid gametes is still speculative, but may be analogous to the process suggested for malaria and coccidia in which the first post-zygotic division represents meiosis (Canning & Anwar, 1968).

Downs (1941) showed that it was possible to establish malaria infection from a single erythrocytic parasite. Using a similar technique, Walliker and colleagues established pure clones of infectious blood forms of various rodent malarias, characterized by specific isoenzyme or drug resistance markers (Walliker, 1976; Carter & Walliker, 1977; Carter, 1978). They then induced controlled matings by allowing mosquitoes to feed on rodents infected simultaneously with parasites of 2 specific clones. The zygotes which developed were allowed to mature within the mosquitoes to produce sporozoites which were used to infect rodents. The infections thus established were cloned and examined for the specific markers. Two important points emerged. The first was that the products of a cross possessed only 1 parental form of each specific marker, in no clones were diploid forms of the same marker found, and since each marker was located at a specific gene locus this meant that reduction division or meiosis had occurred before blood infections were established, in other words, within the mosquito. The second point that emerged was that recombination of different markers could occur and true hybrids could be formed. If, for example, a clone with 2 different isoenzymes A_1 and B_1 was crossed with a clone containing variant isoenzymes A_2 and B_2, resultant homologous forms ($A_1 B_1$ and $A_2 B_2$) could be isolated and also the hybrid forms $A_1 B_2$ and $A_2 B_1$. The significance of these findings in the evolution of malaria parasites, particularly with regard to the emergence of new or drug-resistant strains, has not been overlooked.

Work on coccidia has followed similar lines and it has been shown that drug resistance factors can be transferred by genetic crossing at zygote formation (Jeffers, 1974; Joyner & Norton, 1975, 1977).

If the situation in *Theileria* is analogous to that in malaria and coccidia, the most likely time that reduction division occurs is during sporogony in the acinar cells of the tick salivary gland. This would mean that sporozoites transmitted to the bovine were haploid.

In malaria and coccidia all stages except the zygote appear to be haploid, and since sexually dimorphic forms can arise in clones derived from single sporozoites or erythrocytic forms (Downs, 1941; Walliker, 1972; Shirley & Millard, 1976), it must follow that 'male' and 'female' differentiation arises phenotypically. The situation in *Theileria* is less clear. Firstly, it is not even known whether the bovine stages are haploid and if so when gamete differentiation occurs. Secondly, no sexual dimorphism has been recognized in the bovid, although comma and ring forms of piroplasms have been thought to represent dimorphic gametocytes. However, Schein *et al.* (1977) could not confirm this suggestion, noting that comma forms were usually digested in the tick gut.

Much of the early literature referred to the macro- and microschizont stages of *Theileria* as agamonts and gamonts respectively, implying that reduction division occurred when gamonts (microschizonts) were formed. Electron microscope studies show that the micromerozoites of microschizonts are formed by budding from the macroschizont residual body (Jarrett & Brocklesby, 1966). It would, therefore, be reasonable to assume that a reduction division could occur at this time. This, however, seems unlikely on the basis of analogy with malaria and coccidia, and since we have taken the stand of bringing the piroplasms and malaria parasites closer together, we feel the assumption that the bovine stages of *Theileria* are entirely haploid will prove to be correct. However, in attempting to unravel the genetics of *Theileria* one must guard against too much simplistic analogy with classical Mendelian genetics and terms such as male, female, haploid, diploid, meiosis, chromosome etc. should be used with caution since, although convenient, they may well be incorrect.

PRACTICAL IMPLICATIONS

An organism of limited reproductive potential which relies purely on mutation to overcome the selection pressures of evolution may well be at a disadvantage in competing with other organisms which have the capacity for sharing and disseminating genetic material. Some form of sexual process is the method adopted by most organisms for exchange of genetic material, and it now appears that this is the method adopted by *Theileria*.

In the field situation there are various strains and species of *Theileria*, transmitted by different species of tick to a variety of mammalian hosts, so that the potential for biological variation under the most restrictive selection system (simple mutation) is considerable, but when the prospect of genetic recombination is added the potential for variation becomes infinitely greater. For example, in the case of the rodent malarial parasite *P.c. chabaudi* from West Africa, 2 isoenzyme types have been characterized: 1 occurring in 3 forms and the other in 4. With random mating 12 combinations of these alleles are possible. From 17 isolates, 9 of the possible combinations were found (Carter & Walliker, 1977), thus demonstrating considerable random mixing within the population. A similar situation is described for *P. falciparum* (Carter & Voller, 1975). This situation seems complex enough but when numerous other characters are added to the gene pool the potential for variation becomes bewildering.

Genetic mixing may well be the key to the complex field situation which prevails in theileriosis, particularly East Coast fever (ECF). Classical ECF caused by *T. parva* has been characterized by Brocklesby, Barnett & Scott (1961) on the basis of a large number of observations on cattle experimentally infected with *T. parva* (Muguga). Such a typical syndrome is seen in the field where a relatively simple epidemiological situation prevails with predominantly single species of vector and host. This situation occurs in the cultivated areas of East Africa with dense cattle populations. Strains isolated from these areas normally have classical characteristics. In other areas, however, where there is extensive overlap of different potential tick and host species, the system may be much more complex. It is from these marginal areas that differing strains are sometimes isolated, and it is tempting to suggest that such areas provide the milieu for mixing of separate strains and the potential emergence of new genetic recombinants. Unless natural selection pressures were very strong these new recombinants would have difficulty in emerging in the wild due to the presence of a multiplicity of recombinants which collectively can be assumed to make up the ECF syndrome. However, under experimental laboratory procedures selection of recombinant clones becomes much more feasible. Wilde (1978) suggested this may be the reason why a number of wild strains apparently alter their characteristics under the isolation and passage conditions currently adopted experimentally. Many examples have shown that when *T. lawrencei*, isolated from buffalo (*Syncerus caffer*), is passaged through cattle it transforms to typical *T. parva*. On a single occasion when transformed *T. lawrencei* was put back into a buffalo and then re-isolated it retained its *T. parva*-like properties, although causing only mild disease in the buffalo (Young, personal communication). Similarly, Brocklesby & Bailey (1968) showed that a mild strain of *T. parva* became virulent after experimental passage through cattle. The results of both groups of studies could be interpreted on the basis of selection of genetic recombinants, but possibly the most convincing evidence is that of Brocklesby (1969) who used small numbers of ticks to passage a single theilerial infection of low piroplasm count through a relatively large number of cattle, and thus un-wittingly closely simulated the selection technique later adopted by Walliker and his colleagues. From this single infection Brocklesby isolated at least 3 clinically distinct syndromes which could suggest that different genetic recombinants or clones were isolated. If this interpretation is correct then new light is shed on other findings by earlier workers. Infection rates in ticks taken from the field can sometimes be very low: the majority of ticks being uninfected, with others showing only single acini infected. If such infections arise from a single zygote, as now seems likely, they will be the product of a single genetic recombination with a specific set of genes which are unlikely to express the complete genetic spectrum of *T. parva*. In such a case, an atypical or mild ECF infection may arise when the infected tick subsequently feeds on a host. The host may recover but still be partially or wholly susceptible to further ECF challenge involving a different spectrum of genetic determinants. Infestations with single ticks have been established experimentally and mild infections have sometimes ensued. Similarly, ticks which have been exposed to adverse physical or chemical conditions, which can affect parasite development, may transmit mild or atypical ECF and cattle can subsequently be susceptible to more virulent challenge.

It is now clear that strains of *T. parva* do exist in the field and that animals can be protected against homologous challenge but will not necessarily survive heterologous challenge. It seems reasonable to suggest, therefore, that different genetic recombinants can be selected under abnormal laboratory or field conditions, that immunity of cattle to homologous challenge with such recombinants ensues, irrespective of the severity of the original attack, but that immunity to heterologous challenge will be dependent more on the number of antigenic determinants in the immunizing dose (i.e. the width of the disease spectrum covered) rather than the size of the dose *per se*. A similar idea has been put forward by Lohr (1978) who talks in terms of immunovariants rather than genetic determinants but nonetheless his basic premise is the same. He points out that a single field isolate is unlikely to contain a complete spectrum of immunovariants and suggests that partial failure of a field trial at Ngong in Kenya, where immunized cattle were exposed to field challenge, could be explained on the basis of the appearance of immunovariants not present in the original isolate with which the cattle were initially immunized. It was suggested that these immunovariants were of *T. lawrencei* origin derived from endemic buffalo. Lohr's hypothesis is supported by work carried out by Young and his colleagues (personal communication) who found that *T. lawrencei* parasites isolated from the same buffalo over several months were not necessarily cross-protective to cattle, indicating that the same buffalo was producing antigenically distinct organisms at different times. It was also shown that the same buffalo could harbour 2 biologically distinct strains of *T. lawrencei* at the same time.

The evidence presented shows that biological variation of some sort occurs in *Theileria*. The mechanism whereby variation takes place, however, is not known but the answer may lie with genetic recombination as we propose. Studies of theilerial strains and variants by earlier workers indicate that parameters such as pathogenicity or cross-protection cannot be used to provide the answer since there are too many uncontrollable and variable factors involved. In the light of recent developments it now seems more rational to look for a solution using techniques of isoenzyme characterization and genetic analysis, as used for malaria and coccidia, or monoclonal antibody characterization as being carried out at this Institute.

VACCINATION

Radley, Brown, Cunningham, Kimber, Musisi, Payne, Purnell, Stagg & Young (1975) introduced the infection and treatment method for vaccination against ECF. They injected cattle with a 'cocktail' containing 3 distinct strains of *Theileria* (2 *T. parva*-like and 1 *T. lawrencei*-like), and to control infection they treated the animals with a tetracycline formulation for the first few days following injection. The animals all recovered and resisted subsequent challenge. A wide spectrum of theilerial strains appears to be covered by this regime and, although it has been suggested that a *T. mutans* component might need to be included in the cocktail, this form of immunization shows considerable promise for field use (Uilenberg, Silayo, Mpangala, Tondeur, Tatchell & Sanga, 1977). How can it be improved?

If the concept of genetic recombination in *Theileria* is correct then there will be a limited genetic pool with a limited number of factors which will determine protective immunity in an exposed population of cattle. Unless mutations intervene, the number of protective determinants will remain constant irrespective of the number of cross-overs and genetic recombinants. It appears that Radley *et al.* (1975) may well have included most of the protective determinants against ECF in their cocktail vaccine, but at the same time a number of undesirable factors were also present. The most important one being the virulence of the material; unless the cattle are protected by chemoprophylaxis the majority would die.

Present evidence suggests that adequate protective vaccination against both ECF and *T. annulata* infections can only be achieved when living parasites become established in the host. Attempts to immunize with killed vaccines may result in high antibody titres but animals invariably succumb to tick challenge. Some of the problems inherent in the use of a live vaccine have already been indicated by the work of Radley and his colleagues. A possible way to overcome these problems is through the selection, isolation and characterization of distinct clones with desired characters, and by appropriate genetic selection and recombination to combine the desired characters and eliminate undesireable traits. It remains to be seen whether these means can be exploited, but at the same time it should be noted that at least 1 strain of *T. parva*, apparently with the desired properties, has already been isolated from the field (Barnett & Brocklesby, 1961). Brocklesby (1978) has urged that such strains be tracked down and captured, so that they can be categorized and exploited in the light of recent knowledge and the development of new techniques.

We are most grateful to the following for helpful discussion, advice and criticism during the preparation of this manuscript: Professor D. W. Brocklesby, Drs C. G. D. Brown, M. P. Cunningham, R. E. Purnell, M. W. Shirley, D. Walliker and A. S. Young.

REFERENCES

BAKER, J. R. (1977). Systematics of parasitic Protozoa. In *Parasitic Protozoa*, vol. I, (ed. J. P. Kreier), pp. 35–56. New York: Academic Press.

BARNETT, S. F. & BROCKLESBY, D. W. (1961). A mild form of East Coast fever. *Veterinary Record* **73**, 43.

BROCKLESBY, D. W. (1969). The lability of a bovine *Theileria* species. *Experimental Parasitology* **25**, 258–64.

BROCKLESBY, D. W. (1978). Recent observations on tick-borne Protozoa. In *Tick-borne Diseases and their Vectors*, (ed. J. K. H. Wilde), pp. 263–286, University of Edinburgh.

BROCKLESBY, D. W. & BAILEY, K. P. (1968). A mild form of East Coast fever (*Theileria parva* infection) becoming virulent on passage through cattle. *British Veterinary Journal* **124**, 236–8.

BROCKLESBY, D. W., BARNETT, S. F. & SCOTT, G. R. (1961). Morbidity and mortality rates in East Coast fever (*Theileria parva* infection) and their application to drug screening procedures. *British Veterinary Journal* **117**, 529–31.

CANNING, E. U. & ANWAR, M. (1968). Studies on meiotic division in coccidial and malarial parasites. *Journal of Protozoology* **15**, 290–8.

CARTER, R. (1978). Studies on enzyme variation in the murine malaria parasites *Plasmodium berghei*, *P. yoelii*, *P. vinckei* and *P. chabaudi* by starch gel electrophoresis. *Parasitology* **76**, 241–67.

CARTER, R. & VOLLER, A. (1975). The distribution of enzyme variation in populations of *Plasmodium falciparum* in Africa. *Transactions of the Royal Society of Tropical Medicine and Hygiene* **69**, 371–6.

CARTER, R. & WALLIKER, D. (1977). Biochemical markers for strain differentiation in malarial parasites. *Bulletin of the World Health Organization* **55**, 339–45.

DOWNS, W. G. (1941). Infections of chickens with single parasites of *P. gallinaceum* Brumpt. *American Journal of Hygiene* **46**, 41–4.

HONIGBERG, B. M., BALAMUTH, W., BOVEE, E. C., CORLISS, J. O., GOJDICS, M., HALL, R. P., KUDO, R. R., LEVINE, N. D., LOEBLICH, A. R., WEISER, J. & WENRICH, D. H. (1964). A revised classification of the Phylum Protozoa. *Journal of Protozoology* **11**, 7–20.

JARRETT, W. F. H. & BROCKLESBY, D. W. (1966). A preliminary electron microscopic study of East Coast fever (*Theileria parva* infection). *Journal of Protozoology* **13**, 301–10.

JEFFERS, T. K. (1974). Genetic transfer of anticoccidial drug resistance in *Eimeria tenella* *Journal of Parasitology* **60**, 900–4.

JOYNER, L. P. & NORTON, C. C. (1975). Transferred drug resistance in *Eimeria maxima*. *Parasitology* **71**, 385–92.

JOYNER, L. P. & NORTON, C. C. (1977). Further observation on the genetic transfer of drug resistance in *Eimeria maxima*. *Parasitology* **74**, 205–13.

KOCH, R. (1906). Beiträge zur Entwicklungsgeschichte der Piroplasmen. *Zeitschrift für Hygiene und Infektionskrankheiten* **54**, 1–9.

LEVINE, N. D. (1971). Taxonomy of the piroplasms. *Transactions of the American Microscopical Society* **90**, 2–33.

LEVINE, N. D. (1973). *Protozoan Parasites of Domestic Animals and of Man.* Minnesota: Burgess.

LOHR, K. F. (1978). A hypothesis on the role of *Theileria lawrencei* in the epizootiology of East Coast fever and on its importance in attempting vaccination. In *Tick-borne Diseases and their Vectors*, (ed. J. K. H. Wilde), pp. 315–17. University of Edinburgh.

MEHLHORN, H. & SCHEIN, E. (1976). Elektronenmikroskopische Untersuchungen an Entwicklungsstadien von *Theileria parva* (Theiler, 1904) im Darm der Übertragerzecke *Hyalomma anatolicum excavatum* (Koch, 1844). *Tropenmedizin und Parasitologie* **27**, 182–91.

MEHLHORN, H. & SCHEIN, E. (1977). Electron microscopic studies of the development of kinetes in *Theileria annulata*. Dschunkowsky & Luhs, 1904 (Sporozoa, Piroplasmea). *Journal of Protozoology* **24**, 249–57.

RADLEY, D. E., BROWN, C. D. G., CUNNINGHAM, M. P., KIMBER, C. D., MUSISI, F. L., PAYNE, R. C., PURNELL, R. E., STAGG, S. M. & YOUNG, A. S. (1975). East Coast fever: chemoprophylactic immunization of cattle using oxytetracycline and a combination of theilerial strains. *Veterinary Parasitology* **1**, 51–60.

RIEK, R. J. (1964). The development of *Babesia bigemina* (Smith & Kilborne, 1893) in the tick *Boophilus microplus* (Canestrini) *Australian Journal of Agricultural Research* **15**, 802–21.

SCHEIN, E. (1975). On the life-cycle of *Theileria annulata* (Dschunkowsky & Luhs, 1904) in the midgut and haemolymph of *Hyalomma anatolicum excavatum* (Koch, 1844). *Zeitschrift für Parasitenkunde* **47**, 165–7.

SCHEIN, E., BUSCHER, G. & FRIEDHOFF, K. T. (1975). Lichtmikroskopische Untersuchungen über die Entwicklung von *Theileria annulata* (Dschunkowsky und Luhs, 1904) in *Hyalomma anatolicum excavatum* (Koch 1844). I Die Entwicklung im Darm vollgesogener Nymphen. *Zeitschrift für Parasitenkunde* **48**, 123–36.

SCHEIN, E., WARNECKE, M. & KIRMSE, P. (1977). Development of *Theileria parva* (Theiler, 1904) in the gut of *Rhipicephalus appendiculatus* (Neumann, 1901). *Parasitology* **75**, 309–16.

SHIRLEY, M. W. & MILLARD, B. J. (1976). Some observations on the sexual differentiation of *Eimeria tenella* using single sporozoite infections in chicken embryos. *Parasitology* **73**, 337–41.

UILENBERG, G., SILAYO, R. S. MPANGALA, C., TONDEUR, W., TATCHELL, R. J. & SANGA, H. J. N. (1977). Studies on Theileridae (Sporozoa) in Tanzania. X. A large scale field trial on immunization against cattle theileriosis. *Tropenmedizin und Parasitologie* **28**, 499–506.

WALLIKER, D. (1972). An infection of *Plasmodium berghei* derived from sporozoites of a single oocyst. *Transactions of the Royal Society of Tropical Medicine and Hygiene* **4**, 543.

WALLIKER, D. (1976). Genetic factors in malaria parasites and their effect on host–parasite relationships. *Symposia of the British Society for Parasitology* **14**, 25–44.

WILDE, J. H. K. (1978). Theileriosis *Report of a Workshop, Nairobi, IDRC, Canada.* (ed. J. B. Henson & M. Campbell), pp. 76–85.

Parasitology (1980), **81**, 221-233

The contribution of *Ascaris lumbricoides* to malnutrition in children

LANI S. STEPHENSON

Division of Nutritional Sciences, Cornell University, Ithaca, New York 14853

(*Accepted* 29 *January* 1980)

SUMMARY

There is a need for more applied and experimental research to determine more precisely the relationships between nutrition, particularly childhood malnutrition, and intestinal parasitic infections, particularly *Ascaris* infection. Nevertheless, it is clear that in certain communities *Ascaris* infection is associated with poor growth in malnourished children and that deworming improves growth. Periodic deworming of children using a mass treatment approach may be needed to control soil-transmitted helminths in areas where parasites and protein energy malnutrition are highly prevalent. The main aims of treatment should be to reduce parasite loads below the level of clinical significance for the individual child and to reduce future environmental contamination with infective faeces for the sake of the community.

Ascariasis has failed to attract the attention of many clinicians, pathologists, and even parasitologists. [This is despite the facts that] *Ascaris lumbricoides* is one of the biggest human parasites, and ascariasis is one of the most common parasitic infections . . .

Prof. Zbigniew S. Pawlowski (1978)
Chief, Clinic of Parasitic Diseases
Medical Academy of Poznań, Poland

INTRODUCTION

Until 1970 there were very few studies reported in the scientific literature which examined the relationship between *Ascaris lumbricoides* infection and malnutrition in children. This situation was surprising since a number of workers in international health had reported their common association (Jelliffe, 1953; World Health Organization, 1967; de Silva, 1957) and swine producers had for decades been promoting the necessity of preventing and treating *Ascaris suum* infection in pigs in order to achieve satisfactory and economic growth rates (Spindler, 1947; Batte, 1974).

Indeed it is not uncommon to encounter the following opinion in the medical literature. 'In non-endemic areas it is justifiable to treat all [intestinal worm] infections however light, while in areas where re-infection is likely to occur, only heavy or moderate infections are worth treating unless simultaneous attempts are made to improve environmental hygiene' (Gilles, 1976). This statement can imply that treatment often does not provide significant health benefits since re-infection is likely to occur. However, a number of studies on *A. lumbricoides* infection in children in developing countries, published within this decade, seriously question the concept that deworming in endemic areas provides few or no measurable public health benefits. It is the purpose of this paper to examine the relationship between *Ascaris* infection and malnutrition with regard to (1) the environment in which most infection occurs, (2) potential nutrition–infection interactions, (3) proven nutrition–infection interactions and (4) recent considerations and developments in the prevention and control of ascariasis. Fruitful areas for further research will also be discussed.

THE ENVIRONMENTAL SETTING

A. lumbricoides infection unquestionably constitutes a public health problem from the point of view of prevalence. Recent global estimates indicate that *Ascaris* infection is one of the most common parasitic infections in the world, with an estimated 986 million persons or about one quarter of the world's population infected (Peters & Gilles, 1977). In some areas, over 90 % of children harbour *Ascaris* infection (World Health Organisation, 1969; de Silva, 1957).

Transmission is maintained through improper disposal of faeces from infected persons which contain potentially infective ova requiring relatively warm temperatures and moisture for development. Having reached the infective stage, eggs of *Ascaris* are capable of surviving considerable cold and dehydration. This means that infections tend to be endemic in poverty areas and in places where sanitation is poor, especially in the warm humid areas of the tropics and sub-tropics. It is improbable that light *Ascaris* infections cause serious disease in otherwise healthy well-nourished persons. However, the vast majority of people infected with intestinal helminths live in developing countries or poverty areas of developed countries where various forms of malnutrition and other diseases are common. There is increasing evidence not only of synergistic relationships between malnutrition and infections but also between one infection and another. In the case of *Ascaris*, infections are most common and heaviest in children of pre-school age who are also most likely to suffer from or succumb to protein energy malnutrition (PEM), gastro-enteritis, respiratory infections, malaria and a variety of other nutritional, infectious and parasitic diseases (World Health Organization, 1965).

Rowe (1976) aptly described the setting in which *Ascaris* infection frequently occurs.

It is difficult for those living in temperate climates with good standards of public health and medical care to realize the impact of disease on rural communities in the tropics. For example, if you happen to be born and grow up in rural Africa, you are liable to harbor four or more different disease-producing organisms simultaneously. And yet as a parent, you must be fit enough to work, or your family will starve. In your village every child at times suffers the paroxysms of malaria fever and you and your wife will mourn the death of one or two children from this disease . . . But lacking effective remedies, you tend to philosophize in the **face of** sickness. You make the effort to walk the ten miles to the nearest dispensary when you or your child is ill, but there may be no remedies, and it may be too late . . .

We must keep in mind this environmental setting when considering whether or not *Ascaris* may constitute a significant public health problem.

POTENTIAL *ASCARIS*-NUTRITION INTERACTIONS AND METHODOLOGICAL PROBLEMS

Adult *Ascaris* live in an excellent physiological position for interaction with the nutritional status of their hosts since they are commonly located in the jejunum of the small intestine where most nutrient digestion and absorption take place and where any antigenic components and toxic or bioactive compounds produced by the worms may be easily absorbed. Larval stages of *Ascaris*, during migration through the liver and lungs, may also be able to interfere with the nutritional status. Unfortunately, the nutritional effects of larval *Ascaris* infection in humans have apparently scarcely been studied and cannot be discussed in detail.

Intestinal *Ascaris* may directly affect the nutritional status of a child basically by causing either a decrease in nutrient intake or a functional increase in the body's nutrient requirements. Any condition causing anorexia, such as general malaise due to larval migration through the tissues or intestinal upset due to the presence of adult parasites in the small intestine, would decrease total nutrient consumption. Functional increases in nutrient requirements might occur due to a number of factors such as (a) interference with absorption at the brush border due to diffuse mucosal damage, physical obstruction by worms, or production of anti-proteolytic substances by worms, (b) loss of macronutrients, fluid and electrolytes in cases of diarrhoea and vomiting and (c) consumption by the worms of nutrients which are needed by the host (Jelliffe, 1953; Woodruff, 1978). These effects would be much more likely to achieve clinical significance when a child is heavily infected with parasites, has a marginal nutrient intake, and/or already has other de-bilitating infections. These interactions are logical from a physiological stand-point, but it has been scientifically complicated and ethically difficult to confirm their occurrence in human studies. Layrisse & Vargas (1975) described the dilemma we face as researchers.

Much has been written on the synergistic interaction between parasitic infection and mal-nutrition in which malnutrition decreases the capacity of the host to invasion of parasites and, vice versa, the effect of parasitic infection influences the host's nutrition. But although this concept of mutual effect seems very logical, it has not as yet received substantial support due in part to the fact that human parasitic infection flourishes mostly in areas where mal-nourishment exists since early life and partly because the experimental studies carried out so far have not been entirely satisfactory.

The difficulties confronted for a perusal study of a single parasitic infection in man should also be taken into account. There are environmental influences which can affect either host or parasites, or both, and which vary according to the ecological area under study. It is also important to bear in mind the difficulty in finding populations infected with a single parasite. In many instances, students are forced to work in multiparasitized areas and make a decision which infection is paramount in inducing malnourishment and rule out other less harmful infections. The literature on this topic reflects pathetically all these inconveniences. Contro-versial results, such as ascribing to a parasite a detrimental nutritional effect by a group of investigators and a non-detrimental effect by others, are not unusual. In other cases, mal-nourishment effects ascribed to a parasite have been denied after more careful studies.

Polyparasitism is an unpleasant fact which both researchers and many unfortunate persons, particularly in the tropics and sub-tropics, must face. It complicates our studies as well as their lives. But we cannot wish it away. I would submit that we dare not try to wish it away, lest we risk confining ourselves to designing *only* those clear cut, well controlled single parasite studies and experiments which produce exciting and statistically significant results but bear little relationship to human reality.

There are 3 basic ways of approaching the study of *Ascaris*–nutrition interactions in children. (1) Well-controlled laboratory animal studies; the weanling pig infected with *A. suum* probably being the most appropriate model (Stephenson, Pond, Nesheim, Krook & Crompton, 1980*a*), (2) hospital-based human clinical studies of selected nutrients such as nitrogen balance and vitamin A absorption and (3) community-based or field studies in which simple parameters, such as weight gain or haemoglobin level, can be measured in an entire population. All 3 types of study have contributed to the knowledge of *Ascaris* infection in children, but each type has limitations regarding which hypotheses can be proved or disproved. For instance, one can determine with *relative* ease, in experimentally infected laboratory animals, whether or not a particular degree of worm infestation inhibits animal growth and to what extent. This does not show that a related parasite will have the same effect on human growth, but it does encourage one to study the problem carefully in humans. Similarly, it is very difficult to prove conclusively in a community study of children exposed to many diseases that *Ascaris* infection alone is responsible for a particular growth deficit. However, careful experimental design in the field, coupled with positive evidence from controlled animal studies in which *Ascaris* was the only infection, help to determine the parasite's contribution to the observed growth deficits.

Thus, just as parasites and malnutrition can act in a synergistic manner, so can a combination of studies. The recent evidence for the contribution of *Ascaris* to malnutrition in children will now be examined in this light.

EVIDENCE FOR THE ROLE OF *ASCARIS* IN CHILD MALNUTRITION

The report of a WHO Expert Committee on the control of ascariasis (1967) cites relationships between *Ascaris* infection and stunting, general under-nutrition, avitaminosis, decreased protein absorption, xerophthalmia and ascorbic acid deficiency. There is a huge literature available on various aspects of ascariasis, but very few well-controlled studies deal with the nutritional effects of *Ascaris* on the human host. The best data available show evidence for decreased growth and nutrient absorption in *A. suum*-infected pigs, disturbances of nitrogen, fat, carbohydrate and vitamin A absorption in hospitalized, infected children and decreased growth in children in infected communities followed by improved growth after deworming.

Studies in pigs with A. suum

Most of the studies of *Ascaris* and nutrition in pigs were designed by scientists interested in economic and efficient swine production. Thus, they are of great practical importance to the swine industry. The effects of larval and adult-stage *Ascaris* infection on weight gain were first studied by Spindler in 1947 in 4 infected, growing pigs and 4 control litter-mates. Pigs were weaned at 8 weeks of age, fed a balanced grain ration and infected with 12 000 *A. suum* ova/day (sufficient to produce severe respiratory distress), for 18 days. At the end of 18 weeks, average daily gain in infected pigs was half that of controls (0·38 versus 0·80 lb/day), and the average daily gain correlated indirectly with the number of worms harboured at necropsy (109 worms, 0·03 kg/day, 39 worms, 0·18 kg/day, 20 worms, 0·21 kg/day, 12 worms, 0·32 kg/day). More recently, Stewart, Johnson & Hale (1972), in a complicated factorial experiment, fed barrow pigs to market weight on a high or low protein diet (18–16–14 % or 15–13–11 %) that either did or did not contain added pyrantel HCl (an anthelmintic drug). They then infected pigs with *Strongyloides ransomi* larvae, *A. suum* ova, and *Oesophagostomum* spp. larvae. The number of *A. suum* ova passed/g of faeces (EPG) after 77 days correlated positively with the average daily feed intake ($r = 0·42$) and with feed intake/kg of body weight ($r = 0·42$) when correlations were adjusted for weights and ages. These correlations were positive but no longer statistically significant when the pigs reached market weight. The number of *A. suum* worms harboured at market weight correlated negatively, but not significantly, with average daily feed intake ($r = -0·11$) and correlated positively, but not significantly, with overall feed intake/kg of gain ($r = 0·10$).

Zimmerman, Spear & Switzer (1973) illustrated somewhat more clearly the growth and feed conversion benefits from prevention of heavy *Ascaris* infection in pigs by feeding pyrantel to 11-week-old pigs receiving either high (16 %) or low (12 %) protein diets and simultaneously exposed for 20 days to ground heavily seeded with *A. suum* ova. After the initial 28-day treatment period, all pigs received a 15 % protein diet to 57 kg and a 13 % protein diet to 98 kg of body weight and none received pyrantel. Their findings showed that addition of pyrantel, which inhibits hatching of *Ascaris* ova in the intestine, caused higher average daily gain and an improved feed to gain ratio in pigs on the high protein diet, compared to pigs on the same diet without pyrantel. This was true during the 28-day exposure to *Ascaris* ova (daily gain: 0·70 versus 0·55 kg/day, feed/gain: 2·33 versus 2·53) and from the outset of the experiment until the pigs reached 98 kg of market weight (daily gain: 0·75 versus 0·68 kg/day, feed/gain: 3·46 versus 3·63). Addition of pyrantel to the low protein diet did not improve average daily gain during either period, or feed/gain ratio during the initial 28-day period. Pigs fed low protein diets and given pyrantel showed slightly better overall feed/gain ratios compared to low protein controls (3·58 versus 3·69).

An earlier study (Zimmerman, Speer, Zimmerman & Switzer, 1971), using similar procedures and pigs of the same age, demonstrated that the presence of pyrantel in the low (11 %) protein diet improved the feed/gain ratio but did not do so in the high (16 %) protein diet. These 4 studies had slightly contradictory findings but indicated that *A. suum* infection in growing pigs is related to lower average daily weight gain and, in some instances, poor efficiency of feed utilization.

Very recently some workers used *A. suum*-infected weanling pigs on low or high

protein diets as experimental models for children and examined nutrient losses
with nitrogen and fat balance techniques. They reported lower nitrogen absorption
and retention, lower fat absorption, decreased lactose tolerance and mucosal
lactase activity and a decreased villus height to crypt depth ratio in infected pigs
on low protein diets compared to uninfected controls (Stephenson, Georgi &
Cleveland, 1977; Stephenson *et al.* 1980*a*; Nesheim & Forsum, 1980). They also
reported large increases in small intestine wet weight (about 50 % increase or more)
which appears to be due primarily to a pronounced hypertrophy of the *tunica
muscularis* surrounding the gut. The mechanisms involved in these changes are
unknown. It is difficult experimentally to achieve uniform worm burdens in pigs,
however, these studies are continuing and will hopefully shed more light on the
quantitative relationships between worm load and nutrient losses (M. C. Nesheim,
personal communication).

Clinical studies of A. lumbricoides

Evidence from clinical studies of *A. lumbricoides* infection in children is accumu-
lating to show lower apparent absorption and retention of protein and lower
apparent absorption of fat and carbohydrate in infected children prior to, as
opposed to after, deworming. In the first of these studies, Venkatachalam &
Patwardhan (1953) found that deworming 9 hospitalized, infected Indian children,
harbouring a mean of 26 adult parasites, significantly reduced the mean faecal
nitrogen excretion/day from 1·32 to 0·76 g. The authors also showed that neither
the improved hospital diet, nor the ova excreted by *Ascaris*, nor the effects of the
drugs used in deworming could have accounted for this change in nitrogen excretion.
They concluded that even moderate burdens of *Ascaris* could be responsible for
nutritionally significant losses of dietary proteins in children receiving diets
marginal in protein content.

Improvement in protein nutrition after deworming was also noted nearly
20 years later by Tripathy, Duque, Bolaños, Lotero & Mayoral (1972) in 5 hospitalized children harbouring a mean of 30 adult *Ascaris* and receiving relatively low
levels of protein in the diet (1·0–1·5 g/kg body wt/day). Faecal nitrogen after
deworming decreased significantly by a mean of 6·5 percentage points of dietary
nitrogen (33·5 % before and 27 % after deworming). Tripathy, Gonzalez, Lotero &
Bolaños (1971) also found, in 3 other *Ascaris*-infected children harbouring over
48 worms each, that the decrease in nitrogen excretion after deworming was
accompanied by an increase in nitrogen retention.

More important from the standpoint of energy balance, Tripathy *et al.* (1972)
found that 4 of 5 infected children exhibited mild to moderate steatorrhea.
Faecal fat excretion decreased in all 5 children after deworming (9·9 versus 2·3 % of
dietary fat). They also found impaired D-xylose absorption in 3 of the 5 children
prior to treatment, which improved slightly in 2 cases immediately after deworming. D-xylose was used as an index of carbohydrate absorption. Decreased
carbohydrate absorption is extremely important in terms of energy intake and
growth since typical diets of children and adults in many developing countries
derive as much as 75 % of their calories from carbohydrate sources. They also

reported generalized mucosal damage in small intestinal biopsies taken prior to deworming and postulated that mucosal damage was responsible for the defects in absorption. The degree of mucosal damage had decreased in biopsies taken after deworming.

On the other hand, Teotia, Teotia, Kunwar & Misra (1969) found that only 1 of 5 *Ascaris*-infected Indian children exhibited fat malabsorption, but the severity of infections was not measured. Also, Bray (1953), in a nitrogen balance study in 4 Nigerian children (ages 7–10 years), found no consistent differences in nitrogen absorption or utilization prior to and after deworming. The children studied exhibited wide variations in severity of *Ascaris* infection, in clinical signs of malnutrition, and 1 subject also harboured *Ancylostoma*.

Recently, Freij, Meeuwisse, Berg, Wall & Gebre-Medhin (1979) studied nitrogen, fat and xylose absorption in 6 piperazine-treated and 7 placebo-treated, *Ascaris*-infected Ethiopian children before and after treatment and reported no significant differences between groups. The mean number of worms harboured was quite low (7 and 6, respectively) compared with previous studies, especially considering the small number of subjects per group.

Some of the above studies have shown relationships between *Ascaris* infection and 1 of the 4 major nutritional deficiency diseases: protein-energy malnutrition. In addition, 3 clinical studies in India have indicated that *Ascaris* infection may interfere with vitamin A absorption and may in this way be a contributory factor in xerophthalmia, with its associated morbidity, mortality and blindness. Mahalanabis, Jalan, Maitra & Agarwal (1976) determined vitamin A absorption in *Ascaris* infected adults and 12 healthy controls. Over 70 % of infected patients had malabsorption of vitamin A, judging from serum levels after a radioactive dose. Also 7 of 23 infected patients who received a D-xylose absorption test had abnormally low values (below 20 % of dose excreted in urine in 5 h). Immediately after deworming, vitamin A absorption improved in 13 of the 14 patients re-tested, and D-xylose absorption increased in all 5 patients re-tested. In a second study by Mahalanabis, Simpson, Chakraborty, Ganguli, Bhattacharjee & Mukherjee (1979), malabsorption of water-miscible vitamin A was found in children with *Ascaris* infection combined with *Giardia lamblia* and in a proportion of children harbouring only *Ascaris* infection. Successful treatment of the 8 children with both infections returned vitamin A absorption to normal levels. Treatment of the 5 children with ascariasis alone improved absorption but differences before and after treatment were not significantly different.

Sivakumar & Reddy (1975) demonstrated that 6 *Ascaris*-infected children absorbed significantly less of a test dose of vitamin A from the gut than did 5 control children (80 versus 99 %). The percentage of the radioactive label excreted in the urine did not vary between groups, hence the infected children retained significantly less vitamin A than did controls (68 versus 82 %). None of the children had steatorrhea, since stool fat content was less than 5 g/24 h. Two of the infected children were re-tested 2 months after deworming; vitamin A absorption improved from 83–86 % to 95–96 %. Only trace amounts of the vitamin A label were found in excreted worms, indicating that sequestering of the vitamin by the worms could not have caused the defect in absorption. The authors concluded that ascariasis may be aggravating vitamin A deficiency in areas where xerophthalmia is common.

Field studies of A. lumbricoides

Three longitudinal studies published within the last 5 years provide strong evidence that periodic deworming of *Ascaris*-infected pre-school children improves growth in areas where protein-energy malnutrition is common. Gupta, Mithal, Arara & Tandon (1977) reported that deworming of malnourished pre-school Indian children every 3 months in villages where *Ascaris* was common resulted in increased weight for age after 8 months of study, compared to non-treated controls. Children 6–48 months old in 2 villages were given tetramisole every 3 months for a year, and control children in 2 other villages received a placebo. Of 15 children whose stools had been positive for *Ascaris* ova, 13 treated children gained over 1 percentage point in wt/age during the year and only 2 lost over 1 percentage point. In the placebo group, 16 children gained but 27 children lost over 1 percentage point. The authors concluded that periodic deworming should be an integral part of a supplementary feeding programme or a child health care package where roundworm is a severe problem. They also pointed out that the 6 tablets of tetramisole needed/child/year cost only about 4 % as much as a year's supply of supplementary food for the same child.

An investigation was undertaken in Kenya in 1975–76 to measure the effects of *Ascaris* infection on growth, nutritional status and health of pre-school age children (Stephenson, Latham, Crompton, Schulpen & Jansen,1979; Stephenson, Crompton, Latham, Schulpen, Nesheim & Jansen, 1980*b*). The study was conducted in 2 Kenyan villages where 186 children aged 12–72 months were examined 3 times at 14-week intervals and anthropometric, clinical and stool examinations were performed. At visit I, 85 % of the children were below 90 % weight for age by the Harvard standard, 27 % had *Ascaris* ova in their stools, and mean anthropometric measurements between infected and control children did not differ significantly. All children received an anthelmintic (levamisole) at visit II; the mean number of worms collected was approximately 7/infected child.

In the 14 weeks between visits I and II (before deworming) children with *Ascaris* ($n = 61$) did not differ from controls ($n = 125$) in weight gain or percentage expected weight gain. In the 14 weeks after deworming (II–III), previously infected children showed significantly higher weight gains (0·7 versus 0·5 kg) and percentage expected weight gain (130 versus 98 %) than controls. This represents a difference of 33 % in expected growth rate/14-week period. Before deworming, triceps skinfold thickness decreased significantly in *Ascaris*-infected children compared with controls (− 1·6 versus 0·3 mm). After deworming, skinfold increased markedly in previously infected children compared with controls (2·0 versus 1·1 mm). A multiple regression analysis showed that *Ascaris* infection was by far the most important variable of 37 possible health, nutritional and socioeconomic variables explaining the decrease in skinfold before and the increase after deworming. Thus, it was concluded that even light *Ascaris* infections may adversely influence nutritional status, and deworming may enhance growth.

Willett, Kilama & Kihamia (1979) have also reported improved growth rates in Tanzanian pre-school children treated with levamisole every 3 months for a year compared with placebo-treated controls. Of 78 children with *Ascaris*-positive stools at baseline, the treated group gained 2·31 kg/year compared to 1·91 kg/year for the placebo group. This is a sizeable difference of 21 percentage points in growth rate/year.

On the other hand, Freij *et al.* (1979) did not find significant differences in monthly weight gain or arm circumference in a study of 24 *Ascaris*-infected Ethiopian pre-school children treated with piperazine compared with 20 placebo-treated controls. The worm load in the treated group (7/child) was comparable to that of the Kenyan study, however, the measurement period was rather short (1 month) and the sample size was much smaller than that of the other studies which yielded positive results. The finding that piperazine-treated children had less morbidity during the study period than untreated children may have important implications for child growth and health and certainly deserves study on a larger scale.

Summary of the evidence

The above-mentioned studies, taken together, show that *Ascaris* infection has been responsible for decreases in growth, nitrogen absorption and retention, fat absorption, D-xylose absorption, lactose tolerance and mucosal lactase activity and has been linked to structural abnormalities of the mucosa of the small intestine in both malnourished pigs and under-nourished children in developing countries. It is also clear that deworming of infected malnourished hosts can result in improved growth rates, in the order of a 20–35 % increase, compared with un-treated infected hosts. The fact that these results have been found, both in controlled experimental studies of pigs with only one important parasitic infection and in clinical and field studies of children with an often undetermined number of multiple infections indicate that the presence of *Ascaris* infection in under-nourished hosts is of genuine public health importance. These new studies also provide a causal explanation for the often-mentioned associations between protein-energy malnutrition and *Ascaris* infection that appear in the older scientific literature.

The physiological mechanisms responsible for these decreases in growth are not fully elucidated. Decreased absorption of all 3 macronutrients (protein, fat and carbohydrate) has been reported in pigs and humans and certainly appears to be involved. A decrease in absorption would cause a functional increase in the body's nutrient requirements, because more food has to be consumed to compensate for that lost due to poor absorption. There are indications that a decrease in nutrient intake due to anorexia is also involved in *Ascaris*-infected pigs (Stephenson *et al.* 1980*a*). This mechanism is being studied more thoroughly (Nesheim & Forsum, 1980) and may be of particular importance for young children in developing countries where the quantity of food available can be limited and children may be weaned on to the adults' bulky diet rich in complex carbohydrate and fibre but of relatively low caloric density for a small child's needs.

RECOMMENDATIONS FOR FURTHER RESEARCH

Important questions to be answered or investigated further in terms of *Ascaris*–nutrition interaction include the following.

(1) To what degree does *Ascaris* infection interfere with nutrient absorption and utilization in malnourished experimental animals and pre-school children? Nutrients of special importance for developing countries include protein, fat, carbohydrate, vitamin A and iron.

(2) To what extent does *Ascaris* infection cause anorexia?

(3) Does *Ascaris* infection cause or aggravate a malabsorption-type lesion in the small intestine?

(4) Does *Ascaris* infection interact with other common gastro-intestinal diseases in a synergistic way?

(5) What is the relationship between worm load and growth deficit in malnourished experimental animals and children?

(6) Do the migrations of *Ascaris* larvae commonly cause or aggravate respiratory symptoms and signs in endemic areas?

(7) What simple methods, other than expensive stool examinations, can be developed to determine the approximate prevalence of *Ascaris* infection in a community? For instance, will a simple count of people who say they have passed worms roughly approximate the present prevalence of infection?

(8) How often does re-infection occur after mass treatment in an endemic community? This may vary strikingly from one environment to another.

(9) What are the optimal dose frequencies for periodic deworming programmes to decrease substantially the *Ascaris* infection rates at lowest cost?

(10) What are the health effects of commonly found polyparasitism (for example, hookworm, *Ascaris*, malaria, gastro-enteritis due to bacteria and viruses) on the human and animal hosts? To what extent are these common diseases synergistic?

These investigations will involve research in animals, clinical and metabolic studies in humans, epidemiological investigations and population experimentation including intervention studies and evaluation of control programmes. Such research deserves a high priority because the disease being investigated is so extremely prevalent and has deleterious effects on growth and health.

One must keep in mind the need to study, in animals or humans, parasite–host relationships as they commonly occur in nature, if one is interested in experimental results of major public health significance for mankind. It does not surprise one that very heavy parasite burdens and/or severe under-nutrition are detrimental to health. Of more importance for further study are the extents to which light or moderately heavy infections, coupled with chronic under-nutrition are detrimental over the long term and thus worthy of large-scale treatment and prevention. These studies are difficult and expensive because they attempt to examine marginal chronic effects, but they do need to be carried out if ministries of health and planning are to launch national health campaigns aimed at the average chronically infected person rather than the rarer acutely and severely ill person. Recent popularization of multivariate data analysis techniques, such as multiple regression and path analysis, will make the task of analysing data from longitudinal studies of polyparasitism easier. Cross-sectional studies of human populations are not likely to detect chronic nutritional effects of lighter loads of parasitic infections.

RECOMMENDATIONS FOR CONTROL OF ASCARIASIS

In the past, treatment of lighter infections of *Ascaris* has been discouraged using the following justifications.

(1) Light loads are 'asymptomatic' (and therefore presumably harmless).

(2) The children will become re-infected anyway.

(3) Environmental hygiene and health education are the true answers to control.

Given the new developments in research on nutritional implications of ascariasis, a review of these arguments is now in order. Field studies in communities of *Ascaris*-infected children have now shown that deworming of infected malnourished children is accompanied by weight gain. While it is true that treated children often become re-infected in endemic areas, the fact that treatment may improve growth, even if only for a short period, may justify treatment particularly in pre-school children who are most vulnerable to growth failure and protein energy malnutrition.

It is true that a rise in the standard of living and significant improvements in environmental hygiene should help to reduce the transmission of soil-transmitted helminths. This has been the gospel put forth for decades. But for millions of people in rural areas and urban slums, environmental sanitation has not improved markedly, and the prevalence of intestinal parasites in many communities has hardly decreased. A recent study in Iran (Arfaa, Sahba, Farahmandian & Jalali, 1977) found that improved sanitation by itself may be unlikely to lower drastically the prevalence of *Ascaris* infection. Three villages in Iran were supplied with either improved sanitation alone (1 latrine/family plus a clean village water supply), treatment alone (piperazine for *Ascaris* infection given every 6 months), or both. Four years later, the reduction in rates of *Ascaris* infection for the different methods were: sanitation, 28%; treatment, 84% and both, 79%. Thus, even after a 4-year period, provision of sanitary facilities alone did little to decrease prevalence of *Ascaris* infection, while treatment decreased prevalence by 79% or more.

Provision and usage of latrines and safe water supplies and health education are of course essential parts of community development and can markedly reduce transmission of many diseases, particularly those due to certain bacteria and viruses. However, in the short term, it is questionable whether increased usage of latrines by itself can be expected to markedly lower the prevalence of *Ascaris* infection. The reasons for this involve prolific ova production by female *Ascaris*, huge economic and time inputs needed to provide sanitary facilities in poor countries, cultural habits of people in poverty areas and normal behaviour of pre-school children who do not see faeces or soil to be potentially infec^· ¯e; they are discussed further elsewhere (Stephenson, 1980).

Thus, despite substantial improvements in education, housing, sanitation and the economic situation in many areas, these improvements are unlikely to occur fast enough to stay or decrease the prevalence of *Ascaris* in most developing areas during the next 10 years. This is especially true if the only weapons used are health education and symptomatic treatment after diagnosis in hospitals and outpatient facilities.

The studies on the nutritional aspects of *Ascaris* infection in Kenyan children quoted here were supported in part by a grant from the World Bank, Washington, D.C. These studies were made possible by the assistance and advice of many persons including Dr M. C. Latham and Dr M. C. Nesheim in the United States, Dr D. W. T. Crompton, Ms S. Arnold, Mr D. Barnard in England, and Dr T. Schulpen, Dr A. A. J. Jansen, Dr M. L. Oduori, Mr H. Kinyanjui, Ms B. Maina, Dr D. Wijers, Dr T. Hanegraaf, Dr A. Voorhoeve, Ms W. van Steenbergen, Dr A. Muller, Ms M. van Rens, Ms S. Lakhani, Mr J. Mbuvi, Mr S. Nzomo, Dr A. Cross and Dr C. Forbes in Kenya. The studies of *Ascaris* infection in malnourished pigs were supported by funds from the Department of Pathology, New York State College of Veterinary Medicine (NIH 5S07-RR05462-16), Department of Animal Science, New York State College of Agriculture and Life Sciences and from the Division of Nutritional Sciences (Hatch Project No. 199440) at Cornell University. The author wishes to thank Dr J. R. Georgi and Mr D. J. Cleveland for help with infection techniques; Mr C. Barlett, Mr E. Walker and Mr D. Kirtland for help with animal care; Mrs M. Yamashita and Mrs T. Hashimoto for preparation of histological specimens; Mr E. Steh for nitrogen determinations; and Dr V. Gemert for advice on statistical analyses. The author is indebted to Dr M. C. Latham and Dr M. C. Nesheim for reviewing the manuscript, and to Ms D. Doty for typing the manuscript.

REFERENCES

ARFAA, F., SAHBA, G. H., FARAHMANDIAN, I. & JALALI, H. (1977). Evaluation of the effect of different methods of control of soil-transmitted helminths in Khuzestan, Southwest Iran. *American Journal of Tropical Medicine and Hygiene* **26**, 230–3.

BATTE, G. (1974). Advances in swine parasitology 1973. In *Proceedings of 22nd Annual Pfizer Research Conference*. Chicago.

BRAY, B. (1953). Nitrogen metabolism in West African children. *Proceedings of the Nutrition Society* **7**, 3–13.

DE SILVA, C. C. (1957). Tropical ascariasis. *Journal of Tropical Pediatrics* **3**, 62–73.

FREIJ, L., MEEUWISSE, G. W., BERG, N. O., WALL, S. & GEBRE-MEDHIN, M. (1979). Ascariasis and malnutrition. A study in urban Ethiopian children. *American Journal of Clinical Nutrition* **32**, 1545–53.

GILLES, H. M. (1976). Diseases of the alimentary system. Treatment of intestinal worms. *British Medical Journal* **2**, 1314–16.

GUPTA, M. C., MITHAL, S., ARARA, K. L. & TANDON, B. N. (1977). Effects of periodic deworming on nutritional status of *Ascaris*-infected pre-school children receiving supplementary food. *Lancet* **3**, 108–10.

JELLIFFE, D. B. (1953). *Ascaris lumbricoides* and malnutrition in tropical children. *Documenta de Medicina Geographica et Tropica* **5**, 314–20.

LAYRISSE, R. & VARGAS, A. (1975). Nutrition and intestinal parasitic infection. *Progress in Food and Nutrition Science* **1**, 645.

MAHALANABIS, D., JALAN, K. N., MAITRA, T. K. & AGARWAL, S. K. (1976). Vitamin A absorption in ascariasis. *American Journal of Clinical Nutrition* **29**, 1372–5.

MAHALANABIS, D., SIMPSON, T. W., CHAKRABORTY, M. L., GANGULI, C., BHATTACHARJEE, A. K. & MUKHERJEE, K. L. (1979). Malabsorption of water miscible vitamin A in children with giardiasis and ascariasis. *American Journal of Clinical Nutrition* **32**, 313–18.

NESHEIM, M. C. & FORSUM, E. (1980). The influence of *Ascaris suum* on protein utilization of malnourished pigs. *Federation Proceedings* **39**, 888.

PAWLOWSKI, Z. S. (1978). Ascariasis. *Clinics in Gastroenterology* **7**(1), 157–78.

PETERS, W. & GILLES, H. M. (1977). *A Color Atlas of Tropical Medicine and Parasitology.* London: Wolfe Medical Publications Ltd.

ROWE, D. (1976). In *Tropical Diseases* (pamphlet), pp. 4–7. World Health Organization, 1211 Geneva 27, Switzerland.

SIVAKUMAR, B. & REDDY, V. (1975). Absorption of vitamin A in children with ascariasis. *Journal of Tropical Medicine and Hygiene* **78**, 114–15.

SPINDLER, L. A. (1947). The effect of experimental infections with ascarids on the growth of pigs. *Helminthological Society of Washington, Proceedings* **14**(2), 58–63.

STEPHENSON, L. S. (1980). Nutritional and economic implications of soil-transmitted helminths with special reference to ascariasis. In *Clinical Disorders in Paediatric Gastroenterology and Nutrition* (ed. F. Lifshitz), New York: Marcell Dekker, (in the Press).

STEPHENSON, L. S., CROMPTON, D. W. T., LATHAM, M. C., SCHULPEN, T. W. J., NESHEIM, M. C. & JANSEN, A. A. J. (1980*b*). Relationships between *Ascaris* infection and growth of malnourished pre-school children in Kenya. *American Journal of Clinical Nutrition* **33**, (in the Press).

STEPHENSON, L. S., GEORGI, J. R. & CLEVELAND, D. J. (1977). Infection of weanling pigs with known numbers of *Ascaris suum* fourth stage larvae. *Cornell Veterinarian* **67**, 92–102.

STEPHENSON, L. S., LATHAM, M. C., CROMPTON, D. W. T., SCHULPEN, T. W. J. & JANSEN, A. A. J. (1979). Nutritional status and stool examinations for parasites in Kenyan pre-school children in Machakos District. *East African Medical Journal* **56**, 1–9.

STEPHENSON, L. S., POND, W. G., NESHEIM, M. C., KROOK, L. P. & CROMPTON, D. W. T. (1980*a*). *Ascaris suum*: nutrient absorption, growth, and intestinal pathology in young pigs experimentally infected with 15 day old larvae. *Experimental Parasitology* **49**, 15–25.

STEWART, T. B., JOHNSON, J. C. & HALE, O. M. (1972). Effects of pyrantel HCl and dietary protein on growing pigs infected in different sequences with *Strongyloides ransomi*, *Ascaris suum*, and *Oesophagostomum* spp. *Journal of Animal Science* **35**, 561–8.

TEOTIA, S. P. S., TEOTIA, M., KUNWAR, K. B. & MISRA, S. S. (1969). Fat malabsorption in intestinal parasitic infections. *Journal of the Indian Medical Association* **53**, 577–82.

TRIPATHY, K., GONZALEZ, F., LOTERO, H & BOLAÑOS, O. (1971). Effects of *Ascaris* infection on human nutrition. *American Journal of Tropical Medicine and Hygiene* **20**, 212–18.

TRIPATHY, K., DUQUE, E., BOLAÑOS, O., LOTERO, H. & MAYORAL, L. G. (1972). Malabsorption syndrome in ascariasis. *American Journal of Clinical Nutrition* **25**, 1276–81.

VENKATACHALAM, P. S. & PATWARDHAN, V. N. (1953). The role of *Ascaris lumbricoides* in the nutrition of the host: effects of ascariasis on digestion of protein. *Transactions of the Royal Society of Tropical Medicine and Hygiene* **47**, 169–75.

WILLETT, W. C., KILAMA, W. L. & KIHAMIA, C. M. (1979). *Ascaris* and growth rates: a randomized trial of treatment. *American Journal of Public Health* **69**, 987–91.

WOODRUFF, A. W. (1978). Pathogenicity of intestinal helminthic infections. *Transactions of the Royal Society of Tropical Medicine and Hygiene* **59**, 585.

WORLD HEALTH ORGANIZATION (1967). Control of ascariasis. Report of a World Health Organization Expert Committee. *WHO Technical Report Series no.* 379.

WORLD HEALTH ORGANIZATION (1965). Nutrition and infection. Report of a World Health Organization Expert Committee. *WHO Technical Report Series no.* 314.

WORLD HEALTH ORGANIZATION (1969). Selected helminthic infections–results of surveys 1963–1968. *WHO Statistics Report* pp. 510–26.

ZIMMERMAN, D. R., SPEER, V. C., ZIMMERMAN, W. & SWITZER, W. P. (1971). Effect of pyrantel salts on *Ascaris suum* infection in growing pigs. *Journal of Animal Science* **32**, 874–8.

ZIMMERMAN, D. R., SPEAR, M. L. & SWITZER, W. P. (1973). Effects of *Mycoplasma hyopneumoniae* infection, pyrantel treatment and protein nutrition on performance of pigs exposed to soil containing *Ascaris suum* ova. *Journal of Animal Science* **36**, 894–7.

Parasitology (1980), **81**, 447–463

With 6 plates and 10 figures in the text

Nematode egg-shells

DAVID WHARTON

*Department of Zoology, University College of Wales,
Penglais, Aberystwyth, Dyfed, SY23 3DA*

(*Accepted* 25 *March* 1980)

INTRODUCTION

The transmission of parasites often involves a high mortality of free-living stages in the environment outside the host. This may be offset by a high biotic potential. In addition, adaptations of nematode eggs and larvae that ensure their survival or increase their chances of infecting a host will reduce the potential wastage rate. Increasing transmission will have an effect equivalent to increasing the fecundity of the parasite and, energetically, may be the more favourable strategy.

It is in this context that the structure and function of the nematode egg-shell should be considered. Its complexity, resistance and variability may be regarded as adaptations that increase the survival of the enclosed embryo or larva in the environment to which they are exposed.

The egg-shell is also responsible for the resistance of the egg stage to the action of nematocides and anthelmintics. Its structure and function has attracted interest in an effort to understand this resistance and to facilitate control.

THE STRUCTURE AND CHEMISTRY OF THE EGG-SHELL

The nematode egg-shell may consist of anything from 1 to 5 layers. The most frequently observed pattern consists of an inner lipid layer, a middle chitinous layer and an outer vitelline layer. These layers are formed endogenously by the fertilized oocyte. In addition, the egg-shell may possess 1 or 2 layers secreted exogenously by the uterine cells, and the egg-shells of some species apparently include layers secreted by the epithelial cells of the ovary (McLaren, 1973; Yuen, 1971).

There is considerable variation in egg-shell structure between different orders of nematodes (Fig. 1; Table 1) and even between different species (Pl. 1). The egg-shell is thus one of the most variable features of nematode anatomy. More information is needed on this variation in shell structure before an attempt can be made to state a basic pattern or to classify the variety of structure found.

The lipid layer

The lipid layer is responsible for the extreme impermeability of some nematode egg-shells. In *Ascaris lumbricoides* it has a unique chemical nature, consisting of

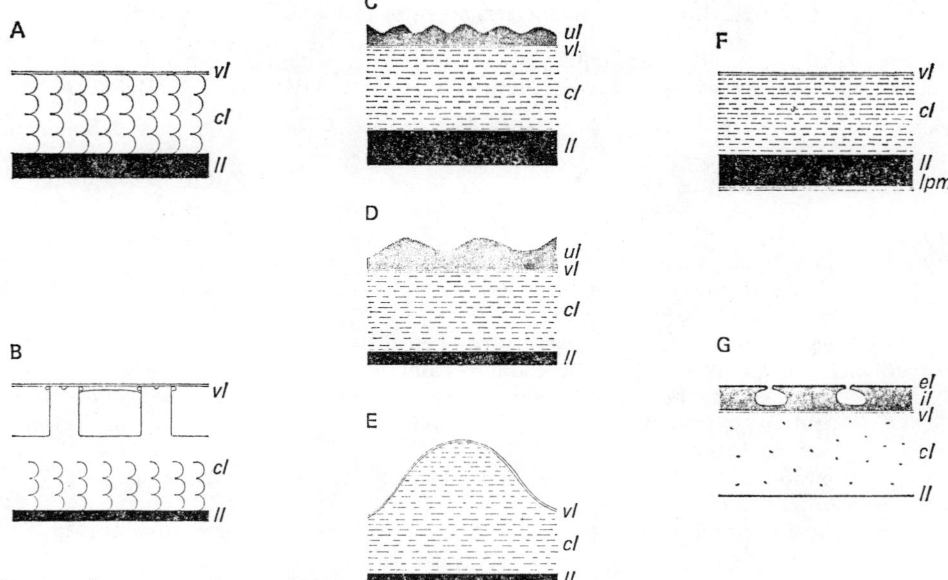

Fig. 1. The structure of nematode egg-shells.

A. *Trichuris suis*: egg-shell consists of an outer vitelline layer (*vl*), a helicoidal chitin/protein chitinous layer (*cl*) and an inner lipid layer (*ll*). (Wharton & Jenkins, 1978).

B. *Capillaria hepatica*: egg-shell consists of an outer vitelline layer (*vl*), chitinous layer (*cl*) and an inner lipid layer (*ll*). The inner portion of the chitinous layer is helicoidal and the outer portion forms a 'pillar and beam-like' network (Grigonis & Solomon, 1976).

C. *Ascaris lumbricoides*: egg-shell consists of an outer uterine layer (*ul*), a vitelline layer (*vl*), a chitinous layer (*cl*) and an inner lipid layer (*ll*). The chitinous layer consists of a chitin/protein complex, the fibres being orientated at random or in parallel. The uterine layer may be acid mucopolysaccharide/protein and is responsible for the mammilated appearance of the egg (Foor, 1967).

D. *Heterakis gallinarum*: egg-shell consists of an outer uterine layer (*ul*), a vitelline layer (*vl*), a chitinous layer (*cl*) and an inner lipid layer (*ll*). The chitinous layer consists of a non-helicoidal chitin/protein complex. The chemical nature of the fibrous uterine layer is unknown (Lee & Leštan, 1971).

E. *Porrocaecum ensicaudatum*: egg-shell consists of an outer vitelline layer (*vl*), a chitinous layer (*cl*) and an inner lipid layer (*ll*). The chitinous layer consists of a non-helicoidal chitin/protein complex and varies in thickness to form a pattern of interconnecting ridges (Wharton, 1979 e).

F. *Tylenchids*: the tylenchid egg-shell consists of an outer vitelline layer (*vl*), a chitinous layer (*cl*) and an inner lipid layer (*ll*). The chitinous layer consists of a non-helicoidal chitin/protein complex. The lipid layer contains a series of lipoprotein membranes (*lpm*). There is considerable variation in the relative thickness of these layers in different species (Bird & McClure, 1976).

G. *Oxyurids*: the oxyurid egg-shell consists of an external uterine layer (*el*), an internal uterine layer (*il*), a vitelline layer (*vl*), a chitinous layer (*cl*) and an inner lipid layer (*ll*). The chitinous layer consists of chitin plus non-protein material. The uterine layers are variously modified to form systems of pores and spaces (Wharton, 1979 b,c,d,).

Table 1. *Variation in the structure of the nematode egg-shell*

Order	Uterine layers	Vitelline layer	Chitinous layer	Lipid layer	References
Trichurida	—*	+	Helicoidal, chitin/protein	+	Grigonis & Solomon (1976) Wharton & Jenkins (1978)
Strongylida	—	+	+	+	Waller (1971)
Oxyurida	Two layers, modified to form systems of pores and spaces	+	Chitin, no protein	+	Wharton (1979 b,c,d)
Ascaridida	Present in some species, may be fibrous, mucopolysaccharidae/ protein	+	Non-helicoidal, chitin/protein, may be modified to form surface pattern	+	Foor (1967) Lee & Leštan (1971) Ubelaker & Allison (1975) Wharton (1979 e)
Spirurida	Absent–surface coat secreted by ovary	+	—	—	McLaren (1973)
Tylenchida	—	+	Non-helicoidal, chitin/protein	Present, plus lipoprotein membranes	Bird & McClure (1976)

*—, Absent; +, present.

25% protein and 75% lipid. The lipid fraction contains a mixture of α-glycosides called ascarosides. These consist of a sugar moiety (glycone), 3,6-dideoxy-L-arabinohexose (ascarylose) and a long chain secondary alcohol (aglycone). Ascarosides can be divided into 3 classes: the monol ascarosides, the diol ascarosides and the diol diascarosides (Fig. 2). In the ovary, ascarosides are present as acetate or propionate esters and are converted into free ascarosides upon the formation of the egg-shell. The material forming the lipid layer may be released at the surface of the fertilized oocyte or by the extrusion of granules stored in the oocyte cytoplasm.

The presence of ascarosides in the lipid layer has led some authors to refer to it as the ascaroside layer (Foor, 1967; Anya, 1976). Ascarosides have been identified in 5 species of ascarid (*Parascaris equorum, Ascaris lumbricoides, A. columnaris, Ascaridia galli, Toxocara cati*) and 1 oxyurid (*Passaluris* sp.). Although it is likely that they are of widespread occurrence, it is perhaps convenient to retain the use of the term 'lipid layer'.

A series of lipoprotein membranes have been described associated with the lipid layer of tylenchid eggs (Bird & McClure, 1976). These have not been observed in the egg-shells of other orders, although this may be due to the difficulty of preparing thin sections for electron microscopy.

The chitinous layer

The chitinous layer is often the thickest layer of the egg-shell and its composition provides structural strength. The presence of chitin has been demonstrated in *A. lumbricoides* by X-ray crystallography (Rudall, 1955) and in *Trichuris suis* by

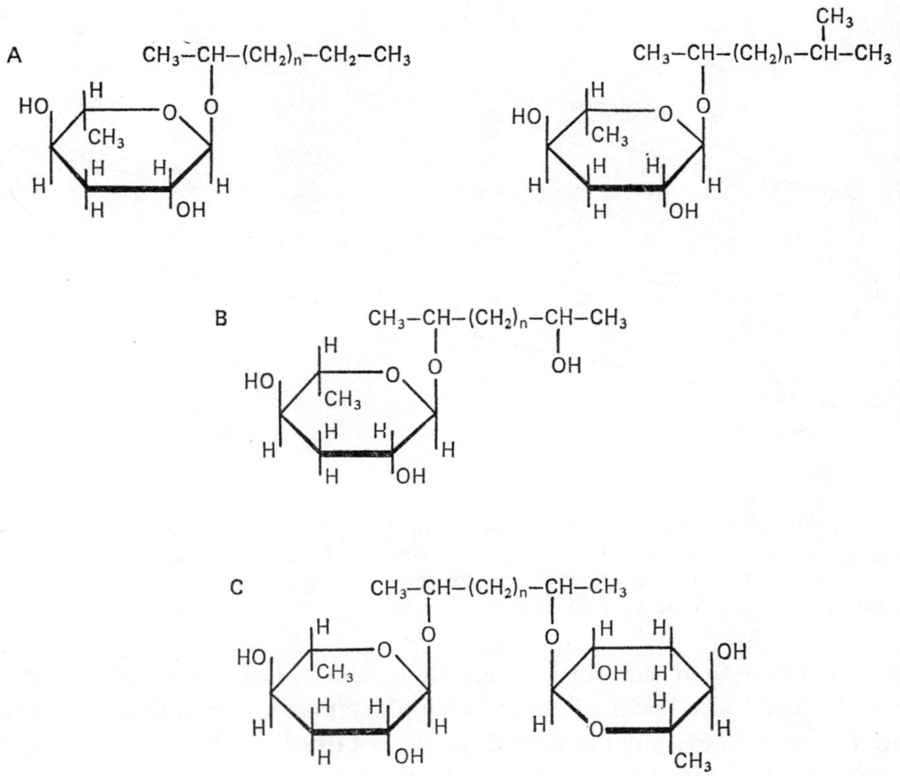

Fig. 2. Ascarosides. The lipid layer contains a mixture of monol ascarosides (A), diol ascarosides (B) and diol diascarosides (C). The chain length of the aglycone is variable with n = an odd number between 21 and 33, depending on the type of ascaroside and the species of nematode (redrawn from Barrett, 1980).

infra-red spectroscopy (Wharton & Jenkins, 1978). Its presence in the egg-shells of other species has been inferred from their histochemistry, solubility properties and the detection of glucosamine in acid hydrolysates. There is no satisfactory histochemical test available for chitin and failure to detect it histochemically or by the chitosan test may be due to the small quantities of material involved or the masking of its reactions by associated proteins. The egg-shell is the only nematode structure in which the presence of chitin has been satisfactorily demonstrated.

Protein is frequently present in association with chitin in the chitinous layer. In oxyurids, however, protein is apparently absent from this layer (Wharton, 1979b,c,d). Chitin/protein complexes frequently occur as 2·8 nm chitin microfibrils embedded in a protein matrix (Neville, 1975). In *T. suis* (Wharton & Jenkins, 1978) and *Porrocaecum ensicaudatum* (Wharton, 1979e) the chitinous layer is made up of fibres which consist of a 2·8 nm chitin microfibril core, surrounded by a protein coat (Pl. 2A and Fig. 3). Examination of the egg-shells of other species may reveal this to be the basic unit of the chitinous layer.

The chitinous layer of *T. suis* has a lamellate structure which in oblique sections appears as bow-shaped arcs (Pl. 2B). This is similar to the structure of arthropod

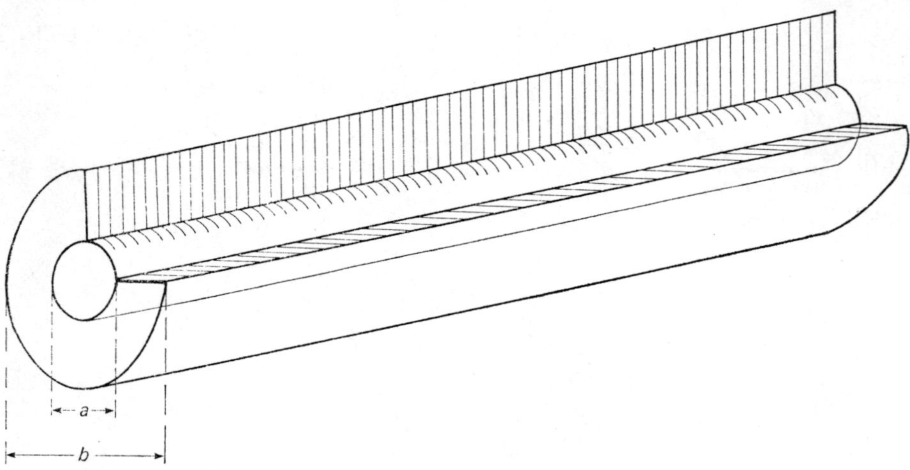

Fig. 3. The basic unit of the chitinous layer appears to be a fibre consisting of a chitin microfibril core (a) surrounded by a protein coat (b). In *Trichuris suis* $a = 2\cdot89$ nm ($\pm 0\cdot578$ nm), $b = 10\cdot1$ nm ($\pm 2\cdot02$ nm). In *Porrocaecum ensicaudatum* $a = 2\cdot95$ nm ($\pm 0\cdot6$ nm), $b = 8\cdot5$ nm ($\pm 1\cdot75$ nm).

cuticle and a variety of biological structural materials. Parabolae and lamellae are artifacts formed by the sectioning of a stack of laminae in which fibres are parallel within any one lamina but the fibre direction rotates in successive laminae. This is known as helicoidal architecture and the formation of this artifact is shown in Fig. 4. A similar structure has been observed in the chitinous layer of *Capillaria hepatica* (Grigonis & Solomon, 1976). There is no indication of helicoidal architecture in the egg-shells of secernentian orders, however, fibres apparently being orientated at random or in parallel (Pl. 2A, C).

The eggs of *P. ensicaudatum* (Pl. 3A), *A. lumbricoides*, *A. suum*, *Toxocara canis* (Pl. 3B) and *T. mystax* possess a surface pattern of interconnecting ridges (Wharton, 1979e; Ubelaker & Allison, 1975). In *P. ensicaudatum* the patterning is formed by a variation in the thickness of the chitinous layer (Pl. 3C).

The vitelline layer

The lipoprotein vitelline layer, derived from the vitelline membrane of the fertilized oocyte, retains a unit membrane-like structure but may become thickened in the fully formed shell. Where the vitelline layer forms the outer layer of the egg-shell, strands of particulate material have been observed adhering to its outer surface (Bird & McClure, 1976; Wharton & Jenkins, 1978; Wharton, 1979e). The origin of this material is unclear (Pl. 4A).

The uterine layers

The outer layers of the egg-shells of some species of nematodes consist of material secreted by the cells of the uterus. Several species of ascarid possess uterine layers, including: *A. lumbricoides* (Pl. 4B), *Heterakis gallinarum* (Pl. 4C) and *Ascaridia galli* (Simonov & Shigina, 1972). Other ascarids do not possess

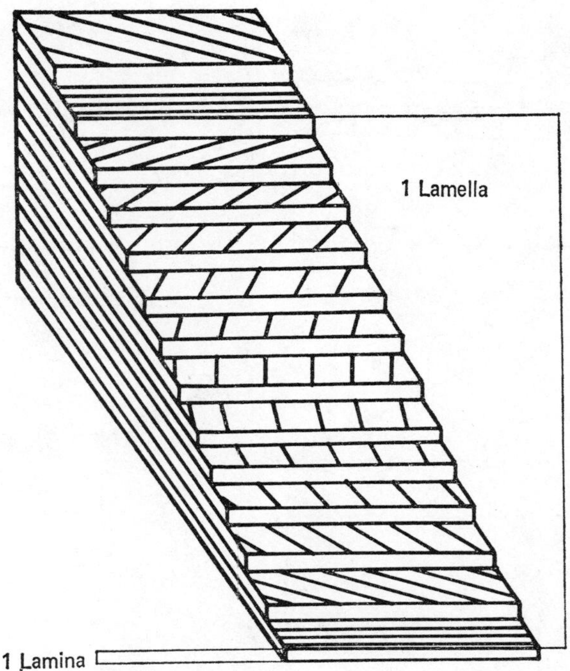

Fig. 4. Diagram showing the Bouligand model of helicoidal architecture. Fibres are parallel in each lamina but fibre direction rotates in successive laminae. Each 180° rotation of fibre direction forms 1 lamella (the boundary of each lamella is where the fibre direction is parallel to the surface of the section – only 1 lamella is shown in the diagram). The fibres of successive laminae trace out parabolic arcs when sectioned obliquely (from Wharton & Jenkins, 1978).

uterine layers (Pl. 3C). The variation in shell structure in the ascaridida requires further investigation and may prove a useful taxonomic feature.

The outer 2 layers of the oxyurid egg-shell are of uterine origin (Wharton, 1979 a,c,d). The lipoprotein external uterine layer consists of a single electron-dense layer with particulate material adhering to its outer surface (Pl. 5A). The internal uterine layer consists of fibrous or granular lipoprotein material and is variously organized to form systems of spaces. In *Aspiculuris tetraptera* the spaces form an interconnecting system (Pl. 5A). In *Hammerschmidtiella diesingi* and *Syphacia obvelata* the spaces appear to be discrete (Pl. 5C, D). The spaces are open to the exterior of the egg via breaks in the external uterine layer. These can be seen to form pores in the surface of the egg-shell (Pl. 6A, B, C). This structure varies in the oxyurids so far examined (Fig. 5) and may prove a useful taxonomic feature.

The complex structures of the uterine layers of oxyurids pose an interesting problem in morphogenesis. There is often no close association between the uterine cells and the egg-shell during the formation of these layers. Their structure does not appear to be formed by a moulding process. During the early stages of their formation, the internal uterine layers of *A. tetraptera* and *S. obvelata* have a crystalline appearance (Wharton, 1979 a,d). The complex structures of the uterine

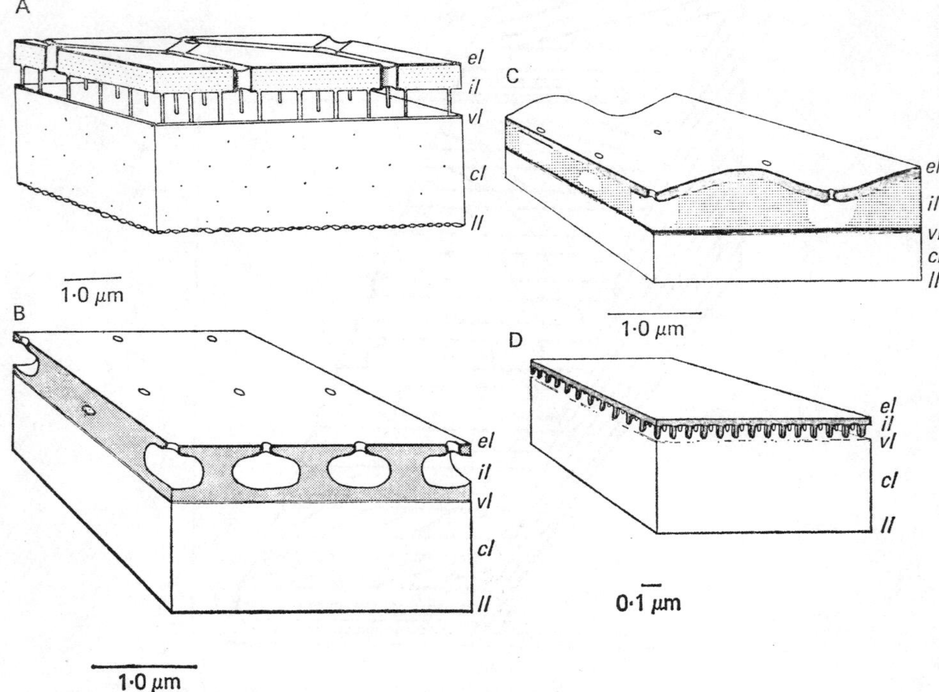

Fig. 5. Structure of the oxyurid egg-shell. The oxyurid egg-shell consists of external uterine layer (*el*), internal uterine layer (*il*), vitelline layer (*vl*), chitinous layer (*cl*) and lipid layer (*ll*). The uterine layers are modified to form systems of pores and spaces.
A, *Aspiculuris tetraptera*; B, *Hammerschmidtiella diesingi*; C, *Syphacia obvelata*; curved side of egg and D, *Syphacia obvelata*; flattened side of egg.

layers of oxyurids may therefore form by a self-assembly process (Wharton, 1979*d*).

Uterine and rectal secretions

Secretions of the vagina uterina and the rectal glands which are associated with egg production have been reported. These include 'sticky coats' around some oxyurid eggs (Lee, 1961; Wharton, 1979*d*) and a gelatinous matrix secreted by the uterine cells or rectal glands of some tylenchids (Bird & Rogers, 1965) and by the uterus of some mermithids (Poinar, 1978).

The operculum

The egg-shells of many species of nematodes possess a specialized region which is associated with the hatching of the larva. The operculum of oxyurid egg-shells consists of a modification of the uterine and chitinous layers (Pl. 6D). In *S. obvelata* the specialization is restricted to the opercular groove which delimits the operculum (Wharton, 1979*d*). In *A. tetraptera* the modification of the egg-shell occurs over the whole of the operculum (Wharton, 1979*b*). The chitinous layer is modified by the presence of electron-dense material, which in *S. obvelata* has been shown to be lipoprotein.

Van der Gulden & Van Aspert-van Erp (1976) have suggested that the hatching of oxyurid eggs involves an increase in the permeability of the egg-shell followed by the opening of the operculum. The specialization of the egg-shell at the operculum may be involved in either or both of these processes.

The trichurid egg-shell is barrel-shaped with an opercular plug at either end. The chitinous layer of the opercular plug consists of a helicoidal chitin/protein complex (Pl. 2 B). There appears to be a higher proportion of chitin to protein in the opercular plug and the arrangement is different from the chitinous layer of the rest of the shell (Wharton & Jenkins, 1978). This may make the opercular plug susceptible to degradation by larval enzymes during the hatching process.

The tanning of the egg-shell

The egg-shells of some species of nematode may be stabilized by a quinone-tanning process. In these species the egg-shell is colourless and soluble in acids, alkalis and various enzymes when dissected from the uterus. When recovered from the host's faeces, however, the egg-shell is brown and insoluble in all re-agents, with the exception of sodium hypochlorite (Monné & Hönig, 1954b). This may indicate the presence of quinone tanning.

Attempts to demonstrate the presence of phenols or phenolase in the reproduc-tive system or egg-shell have frequently been unsuccessful (Anya, 1964; Wharton, unpublished observations). Phenols have, however, been demonstrated in the egg-shell of *Trichuris ovis* (Monné & Hönig, 1954a) and in the ovijector of *Thelastoma bulhöesi* (Lee, 1961). Phenols have also been isolated from the eggs of *A. lumbricoides* by thin layer chromatography after extraction with boiling water (Wharton, unpublished observations).

The evidence is confused and contradictory and the presence of tanning in nematode egg-shells must remain an open question until further information is available.

THE FUNCTION OF THE EGG-SHELL

The nematode egg-shell is one of the most resistant biological structures. It is impermeable to most substances, with the exception of gases and lipid solvents (Arthur & Sanborn, 1969). The penetration of dyes and the effect of temperature on the permeability of the egg-shell suggests that the lipid layer provides the main permeability barrier. The function of the other layers of the egg-shell is less clear.

The main role of the chitinous layer may be to provide structural strength. If it is removed, the lipid layer is easily subjected to mechanical damage and will then allow the entry of harmful chemicals (Arthur & Sanborn, 1969). In most species the chitinous layer consists of a chitin/protein complex. It is thus a composite material consisting of a substance of high tensile strength (chitin) dispersed in the form of fibres in one of lower tensile strength and greater deformability (protein coat). This provides greater structural strength than would either material on their own. The helicoidal arrangement found in the chitinous layers of trichurids and capil-larids makes the egg-shell equally resistant to stress in all directions (Wharton &

Jenkins, 1978). In some ascarids the mechanical resistance of the egg-shell is increased by a system of interconnecting ridges, formed by a variation in the thickness of the chitinous layer (Pl. 3A–C).

Covalent linkages between chitin and protein provide some resistance to chemical attack. This resistance is increased during the browning of the egg-shell in a process which may be akin to the tanning of insect cuticle (Monné & Hönig, 1954*b*). The extreme chemical resistance of the egg-shell is high-lighted by the routine use of 0·1 M sulphuric acid as an embryonating medium and that the action of concentrated sulphuric acid results only in a slight swelling of the egg-shell of *Trichuris suis* (Wharton & Jenkins, 1978).

The permeability of the egg-shell

The impermeability of the egg-shell to water-soluble compounds is indicated by the development of eggs in a variety of harmful solutions. Eggs also develop normally in hyposmotic or hyperosmotic solutions, suggesting that the egg-shell is impermeable to liquid water (Arthur & Sanborn, 1969). However, recent work on the water content of *Ascaris* larvae within the egg in hyperosmotic solutions, indicates that the egg-shell does allow the passage of liquid water (Clarke & Perry, 1980). The eggs of *A. lumbricoides* exposed to desiccation, lose water at a rate dependent upon the relative humidity and temperature; indicating that the egg-shell has a low permeability to water vapour (Wharton, 1979*f*).

The development of the embryo and larva within the egg is aerobic, requiring an exogenous supply of oxygen and the permeability of the shell to oxygen (Passey & Fairbain, 1955). The oxygen molecule is larger than the water molecule; it is therefore not possible to have a biological membrane that is permeable to oxygen but impermeable to water vapour (Hinton, 1969). The egg-shell must therefore have a low permeability to water vapour, even though this may expose the enclosed embryo or larva to the hazards of desiccation.

The nematode egg is small with a large surface to volume ratio. There is a large surface area/unit volume available for the diffusion of oxygen into the egg and, conversely, a large area/unit volume available for the loss of water from the egg. Diffusion is dependent upon the area available and the permeability of the membrane across which diffusion occurs. The egg-shell may therefore have a very low permeability to gaseous exchange; restricting water loss whilst still allowing an adequate supply of oxygen for embryonic development.

The uterine layers of oxyurid eggs

The structural complexity of the uterine layers of oxyurid eggs suggests that they are adaptations which aid the survival of the enclosed embryo or larva. Despite the morphological similarity of these structures to the plastron networks of the chorion of some insect eggs, they do not trap a layer of air when immersed in water and do not function as a plastron network (Wharton, 1980).

When exposed to desiccation the eggs of *H. diesingi* and *A. tetraptera* lose water and collapse at a rate dependent upon the relative humidity and temperature to

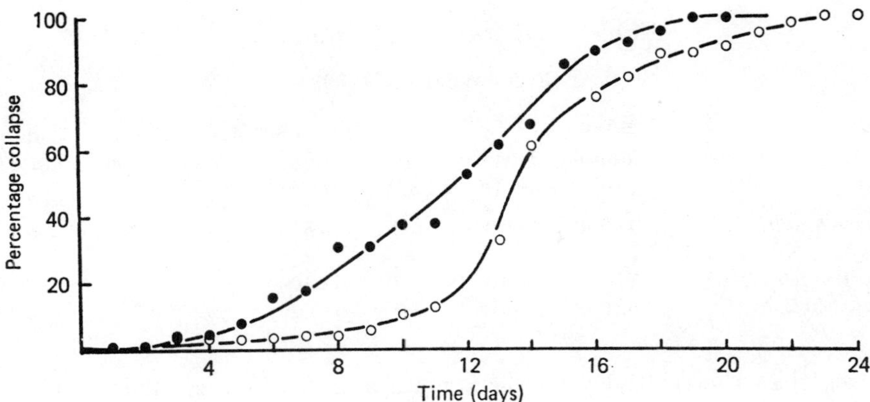

Fig. 6. Collapse of the eggs of *Hammerschmidtiella diesingi* exposed to desiccation at 76 % rel. hum., 20 °C (●) and 16·5 °C (○) (from Wharton, 1980).

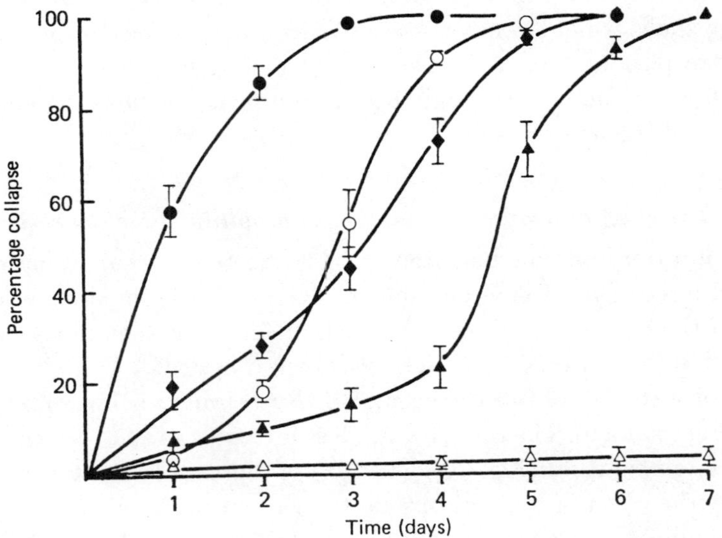

Fig. 7. Collapse of the eggs of *Aspiculuris tetraptera* exposed to desiccation at 20 °C, 0 % rel. hum. (●); 33 % rel. hum. (○); 54·5 % rel. hum. (◆); 76 % rel. hum. (▲) and 98 % rel. hum. (△) (from Wharton, 1980).

which they are exposed (Figs 6 and 7). The egg-shell thus allows a slow rate of water loss from the egg. The rate of diffusion is dependent upon the concentration gradient across the egg-shell, its permeability, thickness and the area across which diffusion can occur (Fick's Law). In addition to restricting permeability, water loss can be controlled by reducing the area available for gaseous exchange.

It has been suggested that this is the function of the uterine layers of oxyurids (Wharton, 1980). If gaseous exchange is restricted to the pores in the external uterine layers there is a reduction in the area available of 85·5 % in *H. diesingi*, 95·6 % in *A. tetraptera* and 96·8 % in *S. obvelata* (Table 2). The complex structures of the uterine layers of *A. tetraptera* and *S. obvelata* result in a much smaller total

Table 2. Dimensions of oxyurid eggs

(From Wharton, 1980)

Species	Volume (μm³)	Surface area (μm²)	S/V	Total pore area (μm²)	Reduction in surface area (%)	Reduction in S/V (%)
Hammerschmidtiella diesingi	48 700	309 900	6·36	44 800	85·5	85·5
Aspiculuris tetraptera	77 900	419 800	5·39	18 400	95·6	95·6
Syphacia obvelata	82 600	386 000	4·67	12 300	96·8	96·8

pore area than the relatively simple structure of *H. diesingi*. This may confirm that the evolutionary trend is towards a reduction in the area across which exchange can occur.

The structures of the uterine layers of oxyurids are similar to the pores in the egg-shells of birds (Rahn, Ar & Paganelli, 1979) and the aeropyles of the chorion of insect egg-shells which do not possess plastron networks (Hinton, 1969). These structures are part of the mechanism by which the egg-shell regulates gaseous exchange between the embryo and the environment and may be considered an example of convergent evolution.

The effect of temperature on the permeability of the egg-shell

The rate of water loss from nematode eggs is affected by changes in temperature encountered in their natural environment. The effect of temperature on the permeability of the egg-shell is thus an important factor in the ability of nematode eggs to survive the hazards of the terrestrial environment.

The rate of water loss from the eggs of *A. lumbricoides* is higher at 30 °C than at 16·5 °C (Figs 8 and 9). The rate of water loss increases as an exponential function of increasing temperature (Fig. 10). The rate of increase is greater than could be explained by the effect of temperature on the vapour pressure of water (Wharton, 1979*f*). This indicates an effect of temperature on the permeability of the egg-shell.

The temperature above which the permeability barrier to dyes and fixatives appears to break down depends upon the time of exposure and whether the eggs are allowed to cool before exposure to the dye (Table 3). Temperature similarly affects the ability of the egg-shell to slow down the rate of water loss (Table 3).

These observations suggest that we are not dealing with a simple 'critical' temperature phenomenon, as observed in insect cuticle, but with the gradual melting or transition of the complex mixture of components that make up the lipid layer of the egg-shell.

Desiccation survival

The eggs of several species of nematodes are able to survive prolonged periods of desiccation (Table 4). This is not due to the prevention of water loss. The structural and chemical properties of the egg-shell do, however, result in a very slow rate of water loss.

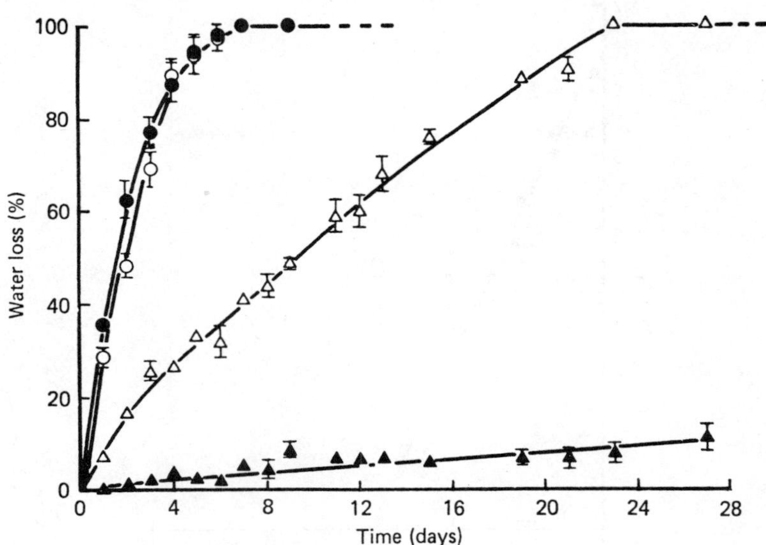

Fig. 8. Rate of water loss from embryonating eggs of *Ascaris lumbricoides* exposed to desiccation at 30 °C; 0 % rel. hum. (●); 33–34 % rel. hum. (○); 75·5 % rel. hum. (△) and 96·5 % rel. hum. (▲). Each point represents the mean ± s.e.m. (calculated from the rate of weight loss, $n = 3$). Taken from Wharton (1979*f*).

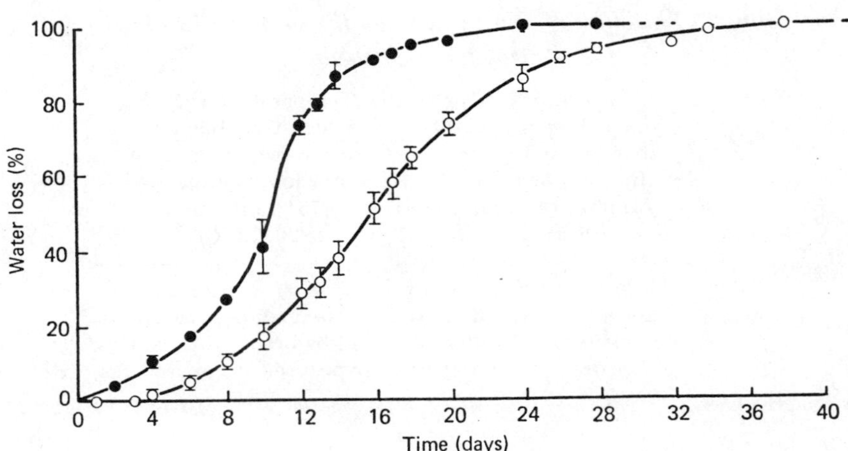

Fig. 9. Rate of water loss from embryonating eggs of *Ascaris lumbricoides* exposed to desiccation at 16·5 °C; 0 % rel. hum. (●); and 32·5 % rel. hum. (○). Each point represents the mean ± s.e.m. (calculated from the rate of weight loss, $n = 3$). Taken from Wharton (1979*f*).

It has been suggested that the slow rate of water loss from the eggs of *Globodera rostochiensis* is achieved by a decrease in the permeability of the egg-shell as it dries (Ellenby, 1968). The mechanism by which this occurs and the layers of the egg-shell involved are unknown.

A slow rate of water loss has been shown to be an essential feature of the ability of some animals to survive anhydrobiotically (Crowe & Cooper, 1971; Evans &

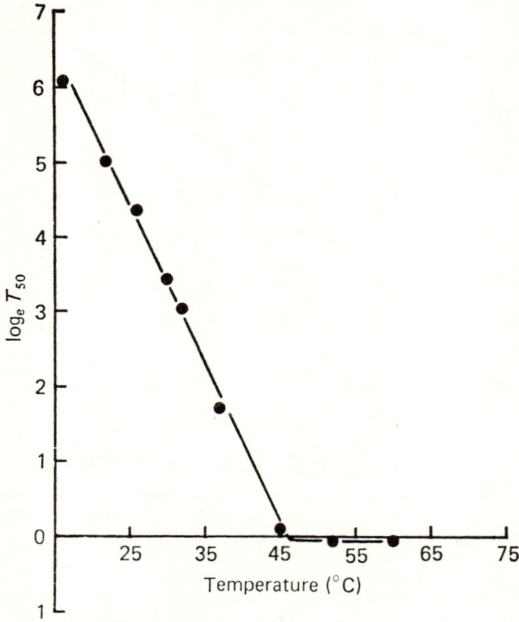

Fig. 10. The effect of the temperature of exposure to 0 % rel. hum. on the time taken for embryonating eggs of *Ascaris lumbricoides* to lose 50 % of their water content (T_{50}). Taken from Wharton (1980).

Table 3. *The effect of temperature on the permeability barrier of nematode egg-shells*

	Temperature (°C) at which the permeability barrier to dye (acid fuchsin) or fixative (osmium tetroxide) is destroyed		Temperature (°C) at which the ability to slow down the rate of water loss is destroyed (76 % rel. hum. 20/22 °C)		
Species	Allowed to cool before exposure	Exposed to dye/fixative at different temperatures	Allowed to cool before exposure	Exposed to desiccation at different temperatures	References
Ascaris lumbricoides	65	44	63–65	15–55	Barrett (1976) Wharton (1979f)
Meloidogyne javanica	50–70	—	—	—	Bird & McClure (1976)
Hammerschmidtiella diesingi	64–82 (1 h) 46–65 (12 h)	64–82 (1 h) 46–65 (12 h)	68–75	32–34	Wharton (1980)

Table 4. *Desiccation survival of embryonated eggs of nematodes*

Species	Temperature (°C)	Rel. hum. (%)	Time (days)	Survival (%)	References
Trichostrongylus colubriformis	20	56	12·5	44	Waller & Donald (1970)
Nematodirus battus	15	33	105	98·5	Parkin (1976)
Trichostrongylus colubriformis	30	65–75	8	40	Anderson & Levine (1968)
Trichostrongylus retortaeformis	20–26	20	21	26	Prasad (1959)

Perry, 1976). The ability of some plant parasitic nematodes to survive anhydrobiotically is well known. The survival of desiccation by the eggs of some animal-parasitic nematodes indicates that the free-living stages of these parasites can similarly enter into a state of anhydrobiosis. This ability may be a significant feature of the epidemiology of parasitic diseases of domestic animals and man.

I would like to thank the following for permission to reproduce published plates: Dr A. F. Bird, Dr M. A. McClure, Dr T. Jenkins, Dr J. E. Ubelaker, Dr W. E. Foor and Professor D. L. Lee. I would also like to thank Mr G. Lloyd for drawing Fig. 3, Dr J. Barrett for reading the manuscript and the A.R.C. for financial support.

REFERENCES

ANDERSON, F. L. & LEVINE, N. D. (1968). Effect of desiccation on the survival of the free-living stages of *Trichostrongylus colubriformis*. *Journal of Parasitology* **54**, 117–28.

ANYA, A. O. (1964). Studies on the structure of the female reproductive system and egg-shell formation in *Aspiculuris tetraptera* Schulz (Nematoda: Oxyuroidea). *Parasitology* **54**, 699–719.

ANYA, A. O. (1976). Physiological aspects of reproduction in nematodes. *Advances in Parasitology* **14**, 267–351.

ARTHUR, E. J. & SANBORN, R. L. (1969). Osmotic and ionic regulation in nematodes. In *Chemical Zoology*, vol. 3 (ed. M. Florkin and B. T. Scheer). London and New York: Academic Press.

BARRETT, J. (1976). Studies on the induction of permeability in *Ascaris lumbricoides* eggs. *Parasitology* **73**, 109–21.

BARRETT, J. (1980). *Biochemistry of Parasitic Helminths*. London: Macmillans.

BIRD, A. F. & McCLURE, M. A. (1976). The tylenchid (Nematoda) egg-shell: structure, composition and permeability. *Parasitology* **72**, 19–28.

BIRD, A. F. & ROGERS, G. E. (1965). Ultrastructural and histochemical studies of the cells producing the gelatinous matrix in *Meloidogyne*. *Nematologica* **11**, 231–8.

CLARKE, A. J. & PERRY, R. N. (1980). Egg-shell permeability and the hatching of *Ascaris suum*. *Parasitology* **80**, 447–56.

CROWE, J. H. & COOPER, A. E. JR. (1971). Cryptobiosis. *Scientific American* **225**, 30–6.

ELLENBY, C. (1968). Desiccation survival in the plant parasitic nematodes, *Heterodera rostochiensis* Wollenweber and *Ditylenchus dipsaci* (Kuhn) Filipjev. *Proceedings of the Royal Society* B **169**, 203–13.

EVANS, A. F. & PERRY, R. N. (1976). Survival strategies in nematodes. In *The Organisation of Nematodes* (ed. N. A. Croll). London and New York: Academic Press.

FOOR, W. E. (1967). Ultrastructural aspects of oocyte development and shell formation in *Ascaris lumbricoides*. *Journal of Parasitology* **53**, 1245–61.

GRIGONIS, G. J. & SOLOMON, G. B. (1976). *Capillaria hepatica*: fine structure of the egg-shell. *Experimental Parasitology* **40**, 286–97.

HINTON, H. E. (1969). Respiratory systems of insect egg-shells. *Annual Review of Entomology* **14**, 343–68.

LEE, D. L. (1961). Studies on the origin of the sticky coat on the eggs of the nematode *Thelastoma bulhõesi* (Magalhães, 1900). *Parasitology* **51**, 379–84.

LEE, D. L. & LEŠŤAN, P. (1971). Oogenesis and egg-shell formation in *Heterakis gallinarum* (Nematoda). *Journal of Zoology* **164**, 189–96.

McLAREN, D. J. (1973). Oogenesis and fertilisation in *Dipetalonema vitae* (Nematoda: Filarioidea). *Parasitology* **66**, 465–72.

MONNÉ, L. & HÖNIG, G. (1954a). On the properties of the egg envelopes of the parasitic nematodes *Trichuris* and *Capillaria*. *Arkiv für Zoologi* **6**, 559–62.

MONNÉ, L. & HÖNIG, G. (1954b). On the properties of the egg envelopes of various parasitic nematodes. *Arkiv für Zoologi* **7**, 261–72.

NEVILLE, A. C. (1975). *Biology of the Arthropod Cuticle*. Berlin: Springer-Verlag.

PARKIN, J. T. (1976). The effect of moisture supply upon the development and hatching of the eggs of *Nematodirus battus*. *Parasitology* **73**, 343–54.

PASSEY, R. F. & FAIRBAIN, D. (1955). The respiration of *Ascaris lumbricoides* eggs. *Canadian Journal of Biochemistry and Physiology* **33**, 1033–46.

POINAR, G. O. (1978). Deposition of mermithid eggs in a gelatinous matrix. *Nematologica* **24**, 135.

PRASAD, D. (1959). The effects of temperature and humidity on the free-living stages of *Trichostrongylus retortaeformis*. *Canadian Journal of Zoology* **37**, 305–16.

RAHN, H., AR, A. & PAGANELLI, C. V. (1979). How bird eggs breathe. *Scientific American* **240**, 38–47.

RUDALL, K. M. (1955). The distribution of chitin and collagen. *Symposium of the Society of Experimental Biology* **9**, 49–71.

SIMONOV, A. P. & SHIGINA, N. G. (1972). Electron microscopy of the egg membranes of *Ascaridia galli*. *Trudy Vsesoyuznogo Instituta Gel'mintologii im. K.I. Skryabina* **19**, 176–86.

UBELAKER, J. E. & ALLISON, V. F. (1975). Scanning electron microscopy of the eggs of *Ascaris lumbricoides, A. suum, Toxocara canis* and *T. mystax*. *Journal of Parasitology* **61**, 802–7.

VAN DER GULDEN, W. J. I. & VAN ASPERT-VAN ERP, A. J. M. (1976). *Syphacia muris*: water permeability of eggs and its effect on hatching. *Experimental Parasitology* **39**, 40–4.

WALLER, P. J. (1971). Structural differences in the egg envelopes of *Haemonchus contortus* and *Trichostrongylus colubriformis* (Nematoda: Trichostrongylidae). *Parasitology* **62**, 157–60.

WALLER, P. J. & DONALD, A. D. (1970). The response to desiccation of eggs of *Trichostrongylus colubriformis* and *Haemonchus contortus* (Nematoda: Trichostrongylidae). *Parasitology* **61**, 195–204.

WHARTON, D. A. (1979a). Oogenesis and egg-shell formation in *Aspiculuris tetraptera* Schulz (Nematoda: Oxyuroidea). *Parasitology* **78**, 131–43.

WHARTON, D. A. (1979b). The structure of the egg-shell of *Aspiculuris tetraptera* Schulz (Nematoda: Oxyurida). *Parasitology* **78**, 145–54.

WHARTON, D. A. (1979c). The structure and formation of the egg-shell of *Hammerschmidtiella diesingi* Hammerschmidt (Nematoda: Oxyuroidea). *Parasitology* **79**, 1–12.

WHARTON, D. A. (1979d). The structure and formation of the egg-shell of *Syphacia obvelata* Rudolphi (Nematoda: Oxyurida). *Parasitology* **79**, 13–28.

WHARTON, D. A. (1979e). The structure of the egg-shell of *Porrocaecum ensicaudatum* (Nematoda: Ascaridida). *International Journal for Parasitology* **9**, 127–31.

WHARTON, D. A. (1979f). *Ascaris sp.*: water loss during desiccation of embryonating eggs. *Experimental Parasitology* **48**, 398–406.

WHARTON, D. A. (1980). Studies on the function of the oxyurid egg-shell. *Parasitology* **81**, 103–13.

WHARTON, D. A. & JENKINS, T. (1978). Structure and chemistry of the egg-shell of a nematode (*Trichuris suis*). *Tissue and Cell* **10**, 427–40.

YUEN, P. H. (1971). Electron microscope studies on *Aphelenchoides blastophthorus* (Nematoda). II. Oogenesis. *Nematologica* **17**, 13–22.

EXPLANATION OF PLATES

PLATE 1

Transverse sections through the egg-shells of 4 species of tylenchid: (A), *Rotylenchulus reniformis*; (B), *Tylenchulus semipenetrans*; (C), *Pratylenchus minyus* and (D), *Meloidogyne javanica*. The egg-shell consists of an outer vitelline layer (*vl*), with particulate material adhering to its outer surface, a chitinous layer (*cl*) and an inner lipid layer (*ll*). These layers vary in thickness and appearance. Lipoprotein membranes (*lpm*), associated with the lipid layer, also vary in their appearance and relative position. Taken from Bird & McClure (1976).

PLATE 2

A. Transverse section through the chitinous layer of the egg-shell of *Porrocaecum ensicaudatum*. The 8·5 nm fibres of the chitinous layer can be seen to consist of an electron-lucent chitin microfibril core, surrounded by an electron-dense protein coat (arrows). Taken from Wharton (1979e).

B. Transverse section through the collar region of the egg-shell of *Trichuris suis*, showing the vitelline layer (*vl*) and the chitinous layer of the shell (*s*) and the opercular plug (*o*). Note the appearance of lamellae in the shell and parabolae in the shell and the opercular plug. Taken from Wharton & Jenkins (1978).

C. Freeze-etch micrograph of a transverse fracture through the chitinous layer of the egg-shell of *Ascaridia galli*. There is no indication of helicoidal architecture; the fibres of the chitinous layer are apparently orientated at random or in parallel. Circled arrow indicates the direction of shadowing.

PLATE 3

A. Scanning electron micrograph showing the opercular plug (*o*) and the pattern of interconnecting ridges on the surface of the egg-shell of *Porrocaecum ensicaudatum*. Taken from Wharton (1979*e*).

B. Scanning electron micrograph showing the surface patterning of the egg-shell of *Toxocara canis*. Taken from Ubelaker & Allison (1975).

C. Transverse section through the egg-shell of *Porrocaecum ensicaudatum*. The egg-shell consists of an outer vitelline layer (*vl*), with strands of particulate material adhering to its outer surface, a chitinous layer (*cl*) and an inner lipid layer (*ll*). The chitinous layer varies in thickness to form the surface patterning. Taken from Wharton (1979*e*).

PLATE 4

A. Transverse section through the chitinous layer (*cl*) and vitelline layer (*vl*) of the egg-shell of *Trichuris suis*. The vitelline layer has strands of particulate material adhering to its outer surface (*pm*). Taken from Wharton & Jenkins (1978).

B. Transverse section through the egg-shell of *Ascaris lumbricoides*. The egg-shell consists of an inner lipid layer (*ll*) chitinous layer (*cl*), vitelline layer (*vl*) and an outer uterine layer (*ul*). Taken from Foor (1967).

C. Transverse section through the egg-shell of *Heterakis gallinarum*. The egg-shell consists of an inner lipid layer (*ll*), chitinous layer (*cl*), vitelline layer (*vl*) and outer, fibrous uterine layer (*ul*). Taken from Lee & Leštan (1971).

PLATE 5

A. Transverse section through the external uterine layer (*el*) and internal uterine layer (*il*) of the egg-shell of *Aspiculuris tetraptera*. The external uterine layer has strands of particulate material adhering to its outer surface (*pm*). Taken from Wharton (1979*b*).

B. Transverse section through the egg-shell of *Aspiculuris tetraptera*. The egg-shell consists of external uterine layer (*el*), internal uterine layer (*il*), vitelline layer (*vl*), chitinous layer (*cl*) and lipid layer (*ll*). The internal uterine layer is modified to form a system of interconnecting spaces which is open to the exterior of the egg via breaks in the external uterine layer (arrows). Taken from Wharton (1979*b*).

C. Transverse section through the egg-shell of *Hammerschmidtiella diesingi*. The internal uterine layer (*il*) is modified to form discrete spaces which are open to the exterior of the egg via breaks in the external uterine layer (arrows). Abbreviations as for Pl. 5B; taken from Wharton (1979*c*).

D. Transverse section through the egg-shell of *Syphacia obvelata*, on the curved side of the egg. The internal uterine layer (*il*) varies in thickness to form ridges and is modified to form discrete spaces which are open to the exterior of the egg via breaks in the external uterine layer (arrow). Abbreviations as for Pl. 5B; taken from Wharton (1979*d*).

PLATE 6

A. Freeze-etch micrograph of the egg-shell of *Hammerschmidtiella diesingi*. The fracture has revealed the external uterine layer (*el*), internal uterine layer (*il*) and the chitinous layer (*cl*). Note the pores in the external uterine layer. Circled arrow indicates the direction of shadowing. Taken from Wharton (1979*c*).

B. Scanning electron micrograph of the egg of *Hammerschmidtiella diesingi*. The specimen is air dried and has collapsed on one side. Note the pores in the surface of the egg and the opercular groove at one pole (o). Taken from Wharton (1979c).

C. Freeze-etch micrograph of the egg-shell of *Aspiculuris tetraptera*. The fracture has revealed the external uterine layer (el), internal uterine layer (il) and chitinous layer (cl) of the egg-shell. Note the system of interconnecting grooves and the pores in the external uterine layer. Circled arrow indicates the direction of shadowing. Taken from Wharton (1979b).

D. Transverse section through the opercular groove of *Syphacia obvelata*. The opercular groove is a modification of the egg-shell, consisting of a thinning and break in the external (el) and internal (il) uterine layers and a corresponding thickening of the chitinous layer (cl) by electron-dense material. Note also the vitelline layer (vl) and the lipid layer (ll). Taken from Wharton (1979d).

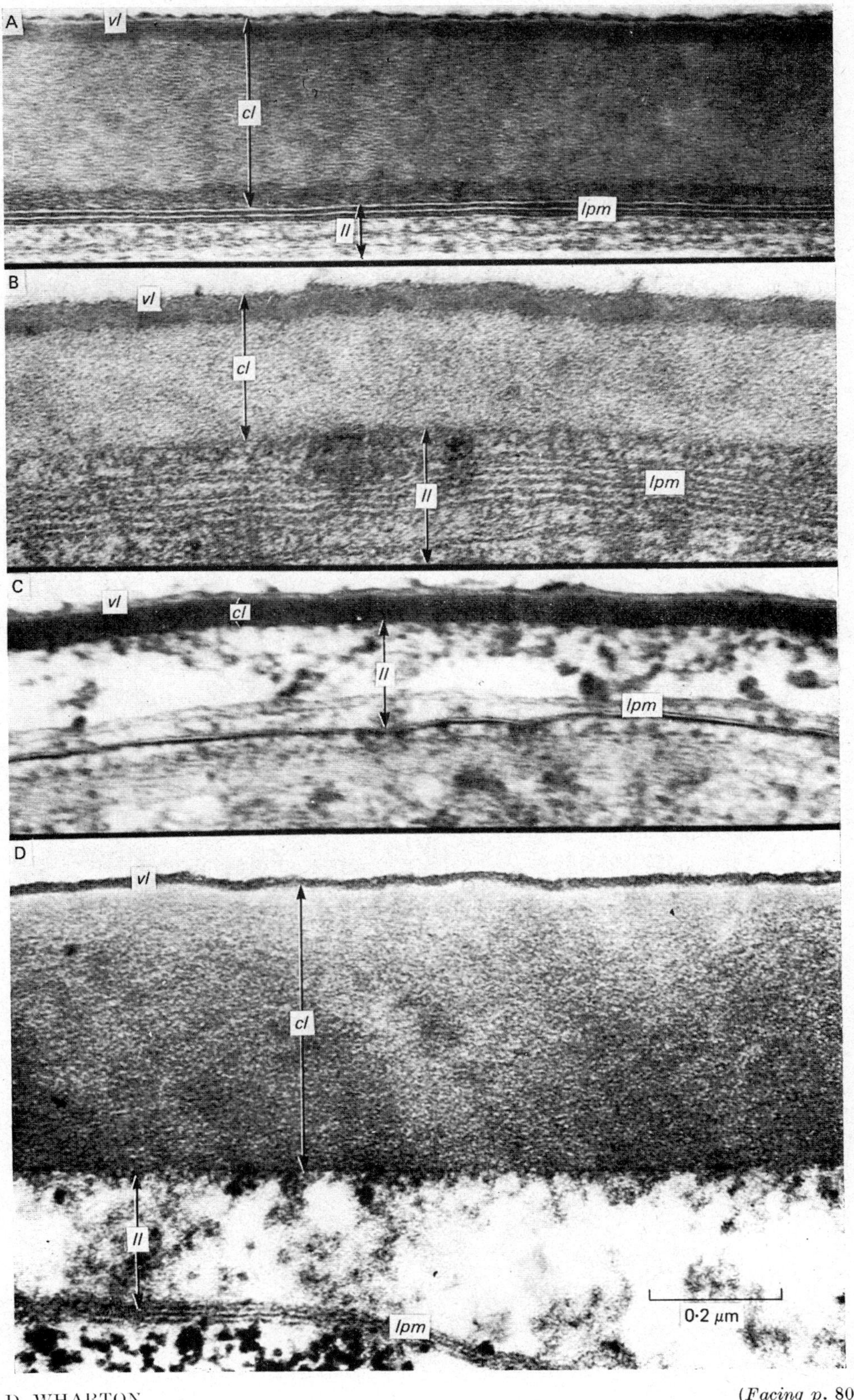

D. WHARTON

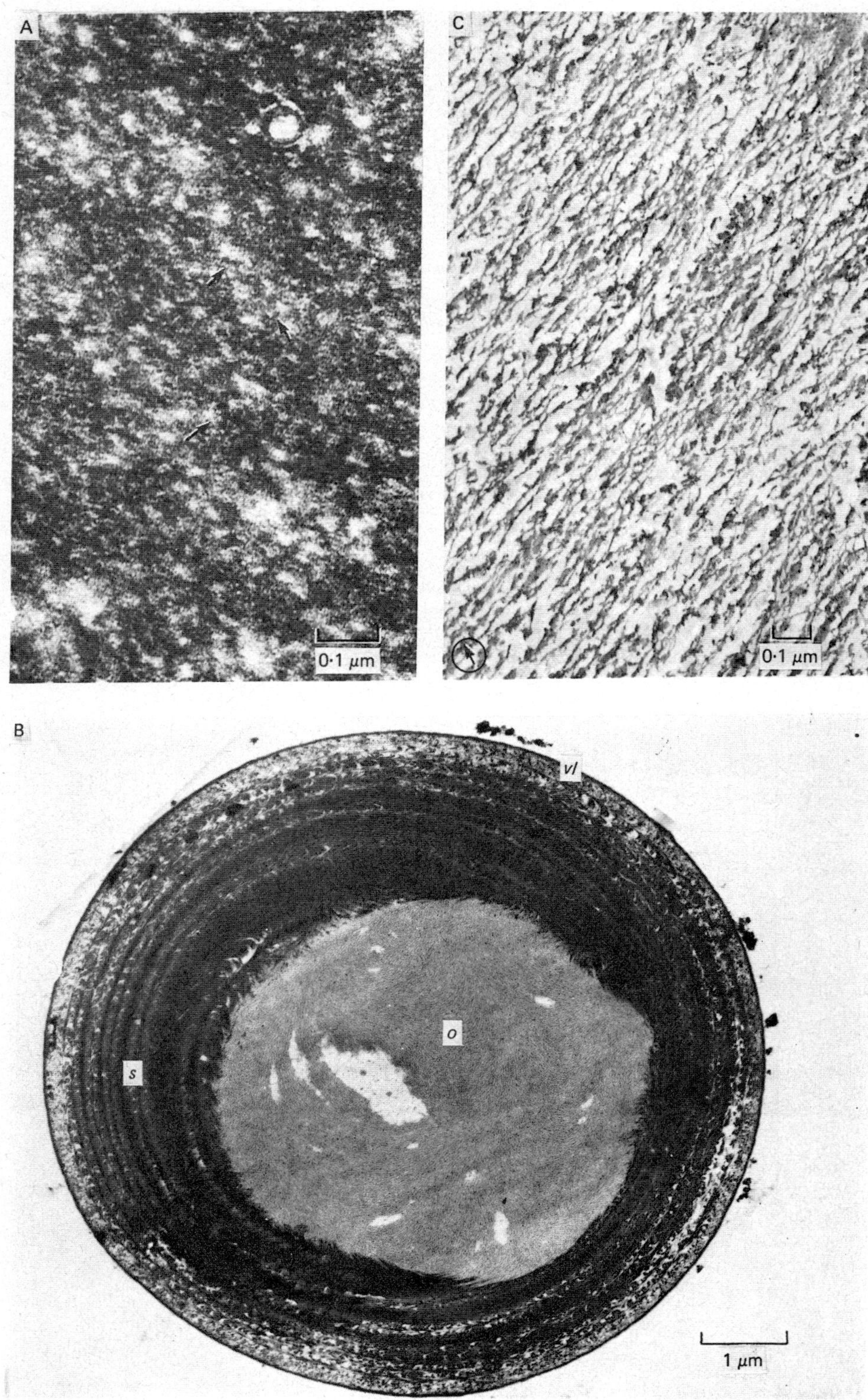

D. WHARTON

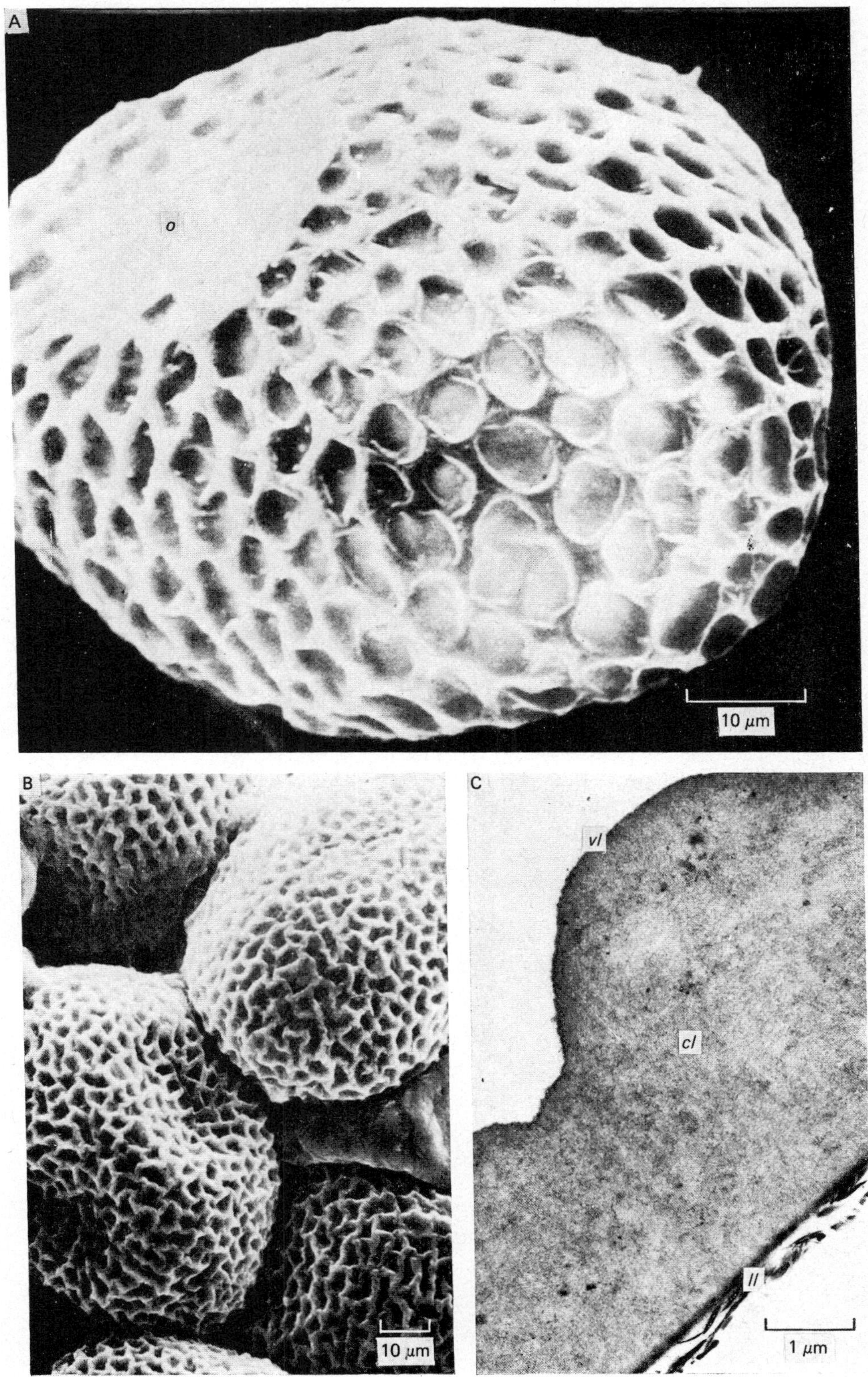

D. WHARTON

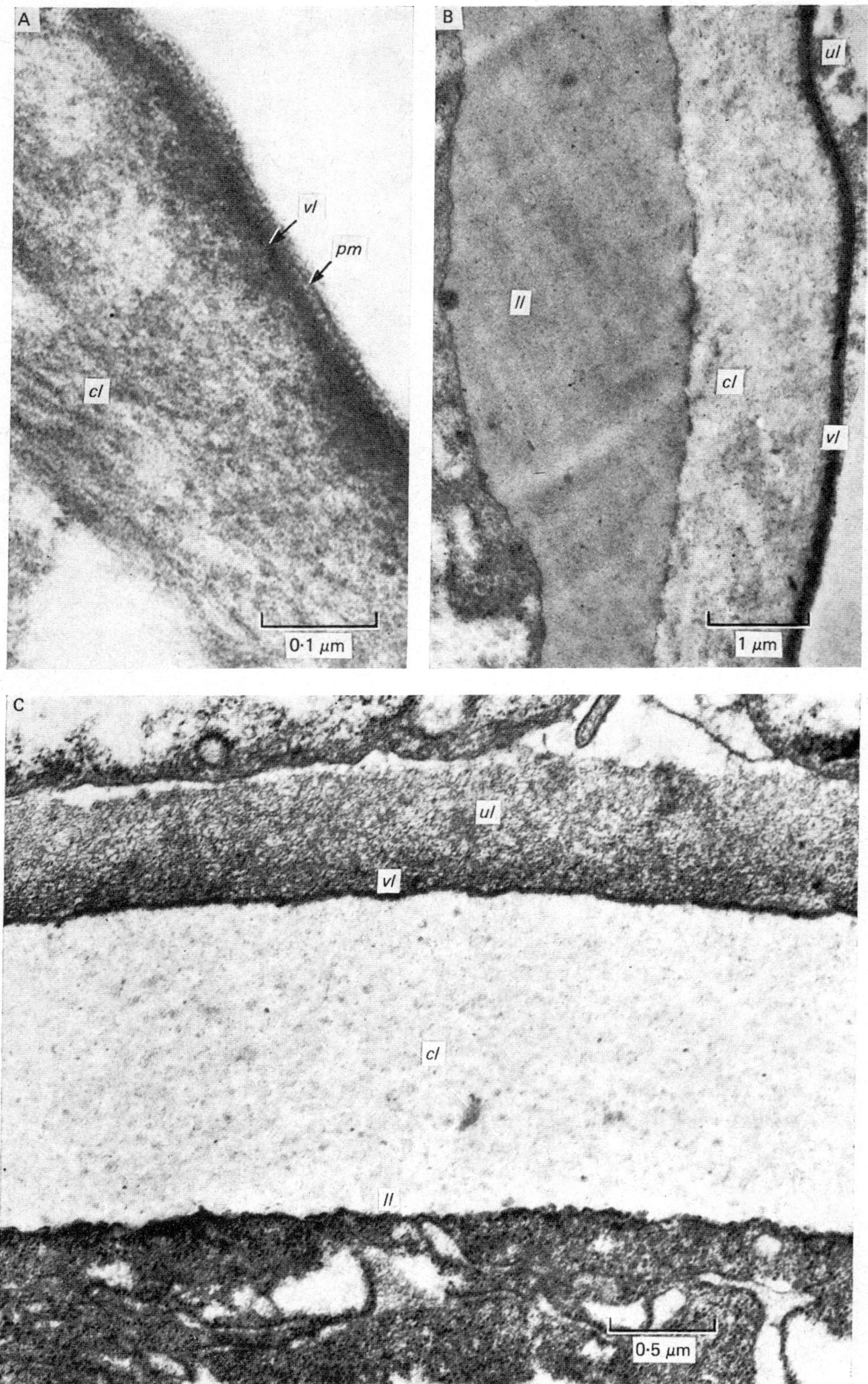

D. WHARTON

B
el
il
vl
cl
ll
1 μm
A
pm
el
il
0·1 μm
C
el
il
vl
cl
ll
0·5 μm
D
el
il
vl
cl
ll
0·5 μm

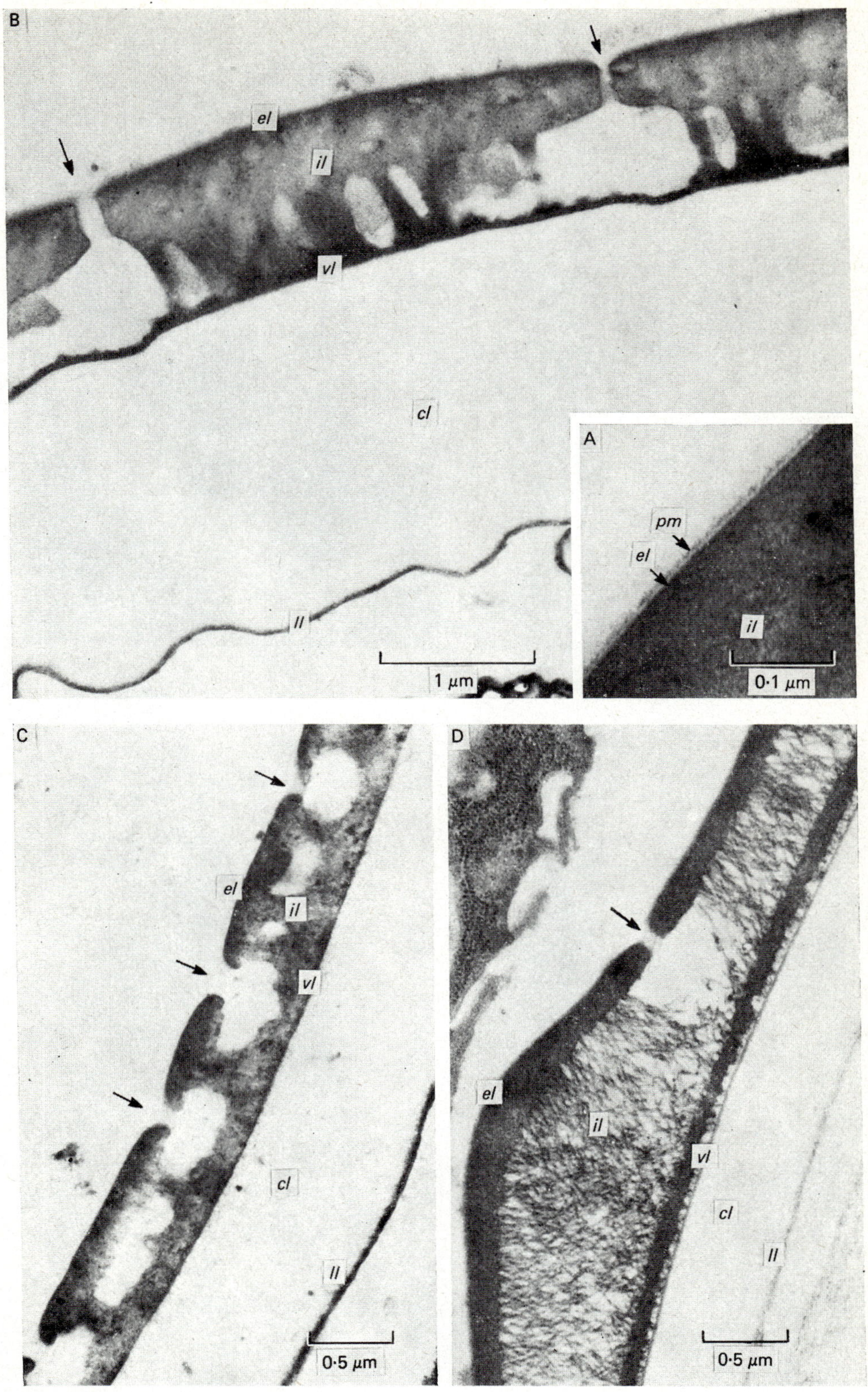

D. WHARTON

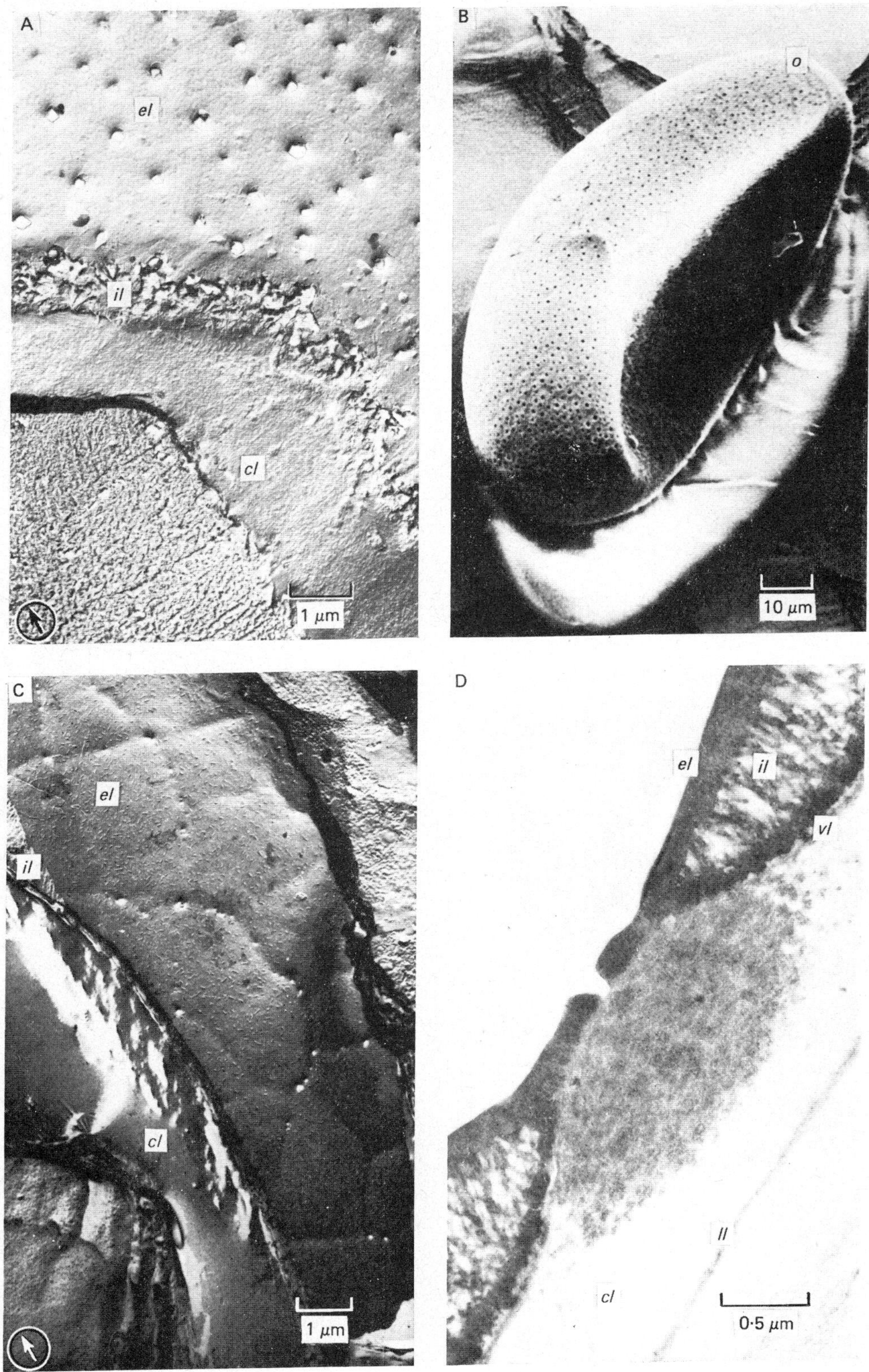

D. WHARTON

Parasitology (1980), **81**, 619–639

With **3** *plates and* **3** *figures in the text*

81

Prospects for the biological control of plant-parasitic nematodes

H. T. TRIBE

Department of Applied Biology, University of Cambridge

(*Accepted* 16 *April* 1980)

Biological control is understood here in the classical sense, which is precisely defined by De Bach (1964) as 'the action of parasites, predators or pathogens in maintaining another organism's population density at a lower average than would occur in their absence'. This account consists of a survey of the principal causes of disease in nematodes, with a summary of the efforts made to use certain pathogens in practise and a discussion of nematode pathology with reference to destruction of plant pathogenic nematodes.

NEMATOPHAGOUS FUNGI

The classical nematophagous or nematode-destroying fungi, pathogenic to vermiform nematodes in general, have been extensively studied. Although a few species were known in the last century, these fungi were brought to prominence by Drechsler at Beltsville, Maryland, in many contributions from 1933 onwards. He termed them 'predaceous fungi' and prepared an early review (Drechsler, 1941). Duddington, in England, reviewed them later (1956) and wrote a popular account (1957). They were the subject of a book in Russian by Soprunov in 1958 which was translated later into English (Soprunov, 1966). Their biology was recently treated in a superbly illustrated monograph by Barron (1977). A key to 97 species was prepared by Cooke & Godfrey (1964) and further species have been described by various authors, notably by Cooke in England and by Barron in Canada, with their co-workers. Nematophagous fungi are generally divided into two biological groups: the nematode-trapping fungi and the endozoic species.

Nematode-trapping fungi

Most of the nematode-trapping fungi operate by formation, on a mycelium, of highly characteristic nematode traps. There are 4 principal types of trap, namely adhesive networks, adhesive knobs, non-constricting rings and constricting rings (Fig. 1). The networks and knobs (Fig. 1*a*, *b*) capture nematodes by a remarkable adhesive, usually secreted in response to contact by a nematode, which holds the nematode firmly to a network or knob cell. After a short interval the fungus grows into the worm and destroys it. Non-constricting rings (Fig. 1*c*) are of a diameter rather less than that of most nematodes. A nematode pushing into the ring, which is typically borne in an upright position on its stalk, becomes wedged and invasion

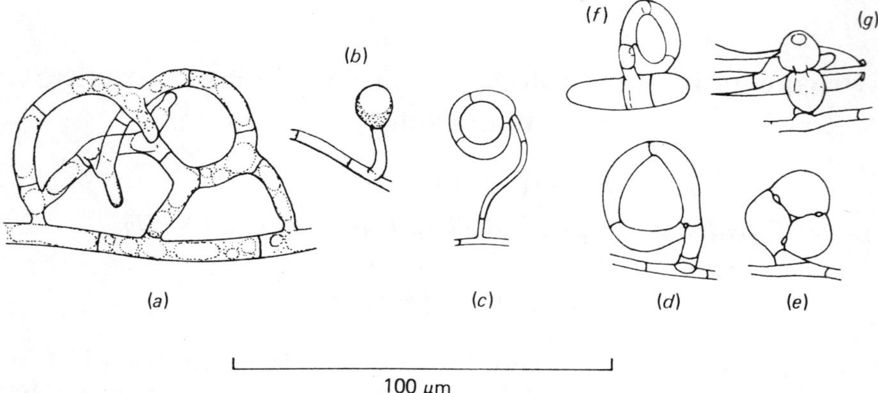

100 μm

Fig. 1. Nematode traps. These are normally borne on the hyphae of nematode-trapping hyphomycetes. (a) Adhesive network; (b) adhesive knob; (c) non-constricting ring; (d) constricting ring; (e) constricting ring, closed; (f) constricting ring arising from a conidium; (g) constricting ring closed over anterior end of a nematode and introducing trophic hyphae into the body of the worm. Drawings made by C. Drechsler (1937) and re-arranged, with acknowledgements to *Mycologia*. Magnification = × 500.

follows. The remarkable constricting ring (Fig. 1 *d–g*), when touched by a nematode pushing into it, snaps forcefully shut and firmly grips the worm, which is duly invaded. Non-constricting rings and adhesive knobs may be pulled off the mycelium by the struggles of captured nematodes, but invasion surely follows from the trapping structure which cannot be dislodged. The fungi which form these rather diverse types of trap are all closely related hyphomycetes, and until about 1964 were placed in the 4 genera *Arthrobotrys, Dactylaria, Dactylella* and *Trichothecium*. Subsequently, several further genera were erected but recently the recognition of biological unity within the group has led to a wholesale transference of species into *Arthrobotrys* (Schenck, Kendrick & Pramer, 1977). Most other species are at present accepted in *Monacrosporium* (which approximates *Dactylella* as understood before 1964), but it is readily arguable that these *Monacrosporium* species are no more distant from the type species of *Arthrobotrys* than are some species subsumed under *Arthrobotrys* by Schenck *et al.* (1977).

Two other groups of nematode-trapping fungi are known which have probably specialized to this activity by independent parallel evolution. One group forms adhesive knobs which stick even more firmly to the nematode than those mentioned above. They also differ in that the knob cell is hour-glass shaped instead of ovoid or globose. The group is represented by species of *Nematoctonus* and has affinities with the basidiomycetes. The second group, in the phycomycetes, does not form trapping organs, but the mycelium secretes adhesive. The genera belonging here are *Stylopage* and *Cystopage*, but most representatives of this group fail to form reproductive structures by which they can be identified and thus can only be classed as sterile adhesive non-septate mycelium. Several studies on relative abundance of nematophagous fungi in soils have shown that this mycelium is very frequent, although *Stylopage* or *Cystopage* species are rarely recorded (Table 3 in

Table 1. *Principal studies on biological control of nematodes
by nematode-trapping fungi*

Principal investigator	Period of publications	Target nematode	Crop or test plants	Location
Linford	1938–9	*Meloidogyne*	Pineapple	Hawaii
Deschiens	1941–3	*Meloidogyne*	Begonia	Paris
Duddington	1956–61	*Heterodera*	Various	England
Hams & Wilkin	1961	*Heterodera*	Various	England
Soprunov	1958	*Meloidogyne*	Various	USSR
Tarjan	1961	*Radolphus*	Citrus	Florida USA
Mankau	1961	*Meloidogyne*	Tomato and okra	California USA
Cayrol & Frankowski	1978–	*Ditylenchus* and *Aphelenchus*; *Meloidogyne*	Mushrooms; various	France

Duddington, 1957; Norton, 1963; Estey & Olthof, 1965; Fowler, 1970). The most abundant nematophagous fungi in soils are network-forming species of *Arthrobotrys*.

The nematode-trapping fungi were used in several attempts to control plant pathogenic nematodes over the period 1938–1961. These investigations are summarized in Table 1; several were quite large-scale and prolonged studies from which numerous papers were published. These attempts have been discussed in the reviews of Duddington (1956, 1957, 1962) and Duddington & Wyborn (1972) and from the bibliographies to these reviews the original papers are accessible. Although some of the earlier studies showed promise in nematode control, the majority produced negative or inconclusive results. No practical treatments utilizing nematode-trapping fungi have stemmed from these studies but practical treatments in horticulture are now being instituted from the recent work of Cayrol & Frankowski (1978, 1979).

Two principles were followed in the early studies. The first was addition to soil of spores or air- or vacuum-dried mycelial preparations of given species and strains of nematode-trapping fungi to boost the low level of populations normally found in soils. The second was addition of decomposable organic matter, usually green plant tissue, to stimulate development of saprophytic nematodes in order to increase the naturally-occurring populations of nematophagous fungi. Sometimes the two principles were combined. Although it appears self-evident that the second principle is valid, R. C. Cooke demonstrated by a number of experiments that it was not so. Cooke prepared microscope slides bearing discs of weakly nutrient agar and buried these in soils amended with chopped green plant tissues and certain other organic amendments. The discs served as indicators on which the number of the several types of trapping organ appearing in the soil environment at weekly intervals after addition of the amendments was estimated. With reference to chopped cabbage leaf tissue added to a medium loam (for example), Cooke showed that there was indeed both a rise in the free-living nematode population and an

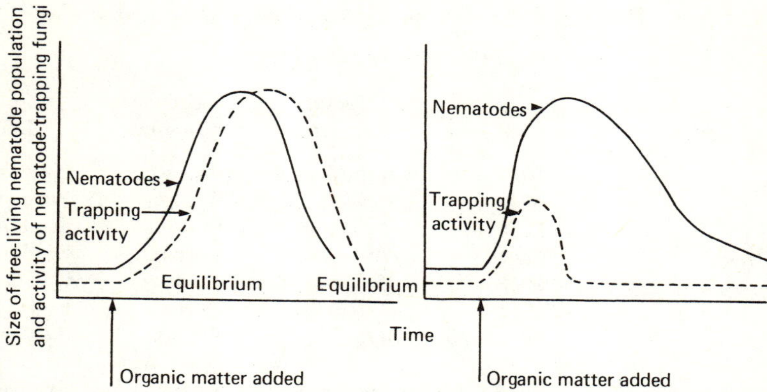

Fig. 2. Relationship between a free-living nematode population in the soil and nematode-trapping fungi after addition of organic matter to the soil: (a) expected changes in nematode population and trapping activity if a labile equilibrium exists between the two groups of organisms; (b) actual pattern of change. From R. C. Cooke (1977) *The Biology of Symbiotic Fungi*, published by John Wiley & Sons, by kind permission.

increase in the activity of nematode-trapping fungi. However, the period of activity was relatively short (4–8 weeks), it did not appear directly related to the density of the nematode population and frequently decreased as the nematode population increased (Cooke, 1968). Later, Cooke (1977) expressed the conclusions from his experiments in graphical form and his graphs are reproduced in Fig. 2.

Cooke (1968) further stated that for predacious activity it was necessary to have organic matter in a particular phase of decomposition, with readily available carbon sources not being exhausted, pointing out that mycelial growth and trap formation are energy-requiring processes that precede predation. Stirling, McKenry & Mankau (1979) have also cited results which they accept as compatible with the theory that nematode-trapping fungi grow saprophytically in the soil and are not dependent on nematodes as a food source. However, many examples are known of conidia of trapping fungi germinating to produce a trap instead of a mycelium (Fig. 1f). There are several illustrations of this phenomenon given by Drechsler (1937) and Figs 45 and 52 in Barron (1977). Barron's view is that the evidence points towards the nematode as the principal, if not the sole, source of organic energy. In soil, a nematode trapped and destroyed by a germinated spore provides a food base very much larger than the conidium. From this base mycelium can spread over soil particles and set more traps. That mycelium of trapping fungi appears independent of nutrient sources other than nematodes is indicated by some incidental observations made by the author during studies of decomposition of cellulose film in soils.

Cellulose film is a thin transparent substrate ideally suited to microscopical observation of the course of its microbial colonization and decomposition (Tribe, 1967). When supported on glass slides and buried in agricultural soil samples it is seen to be colonized by fungi which are succeeded by bacteria. Protozoa and nematodes are predacious on the bacteria and the nematodes are subject to fungal parasitism. In general, some weeks elapse between the establishment of nematode populations and the appearance of nematode-trapping fungi. Network, knob and constricting ring traps and unspecialized sticky mycelium have all been seen and the hyphae bearing these grow as single widely-spaced strands. Nematodes are trapped and destroyed and the hyphae appear quite independent of the background organic matter which usually consists of bacterial populations whose substrate is exhausted or nearly so. Examples are shown in Pls 1 and 2.

The residue from which Pl. 1 has been prepared is derived from cellulose film buried for 80 days in a sample of clay soil. A constricting ring trapping fungus is established on parts of the residue. It has destroyed 13 nematodes, and 115 unsprung rings await further prey. Prey is abundant, since there are 101 healthy nematodes in the vicinity; many are small and there are 25 eggs present, so that nematode reproduction is evidently succeeding. The ring fungus does not appear to be particularly efficient in controlling the nematodes. Although it has almost cleared the nematodes from the area on which it has set the traps the surrounding area is quite thickly populated. It is possible that more than 13 nematodes have been destroyed since destruction under moist, temperate conditions is rapid: a nematode is invaded from the ring cells and 'in a matter of hours, little but the cuticle of a trapped nematode remains' (Duddington, 1957, p. 68). However, there are indications in Pl. 1 that the mycelium is mature and tending to senescence, despite the unconsumed prey. The rings in the slide from which the tracing was made stained very positively with picronigrosin in lactophenol, indicating rich cytoplasm, but the connecting hyphae were mostly poorly stained and appeared moribund. The free ends of the hyphae were not richly cytoplasmic and advancing, but were the remains of hyphae which had lysed away. Examination of another piece of film buried in the same soil sample for 115 days showed a similar picture.

Evidence from several sources therefore indicates that the period of activity of nematode-trapping fungi is limited in habitats within the soil environment, even though the nematode population continues its existence. There is no real evidence of specificity in these fungi; although in agar plate studies quantitative differences in the capacity of given isolates to destroy several species of plant-parasitic nematodes have been demonstrated, the nematode-trapping fungi as a group can destroy virtually any vermiform nematode, human and animal parasites as well as saprophytes and plant parasites (Duddington, 1957). Duddington's analogy of the balance between lion and antelope is probably a good one!

The studies of Cayrol and Frankowski at the Institut National de la Recherche Agronomique in Antibes, begun more than 15 years after the general abandonment of attempts at biological control by nematode-trapping fungi, have their applications in horticulture and market-gardening. The first study concerns the use of a nematode-trapping fungus to protect mushrooms from mycophagous nematodes, principally *Ditylenchus myceliophagus* but also *Aphelenchus avenae*, *Aphelenchoides bicaudatus* and *A. composticola*. A strain of *Arthrobotrys robusta* was used, which in preliminary tests was shown to be entirely compatible in compost culture with the mushroom *Agaricus bisporus*. The *A. robusta*, strain Antipolis, is grown on a rye-based medium and added to the pasteurized mushroom compost at the same time and in the same volume as the mushroom spawn. The *Arthrobotrys* inoculum is produced as 'Royal 300' at the Laboratoire 'Royal Champignon' in St Hilaire-St Florent, France (Cayrol, Frankowski, Laniece, d'Hardemare & Talon, 1978).

A second study (Cayrol & Frankowski, 1979) is directed towards control of *Meloidogyne* in market-garden crops. For this purpose another strain of *Arthrobotrys* is used, strain 1141b, and is commercially prepared as 'Royal 350'. The granular rye-based inoculum is added to *Meloidogyne*-infested soil at the horticultural level of 140 g/square metre and lightly incorporated into the surface 30 days before the plants to be protected are placed in the soil. A crop not susceptible to the nematode is grown during the 30-day stabilization period. Cayrol has stated that this method of biological control is now used in France, with general satisfaction, unless the soil reaction is below pH 6·5, when the *Arthrobotrys* fails to grow in the soil.

Endozoic parasites

Endozoic nematophagous fungi invade the host from adhesive spores which stick on to the cuticle, from spores which are swallowed and lodge in the alimentary canal, from motile spores in water, or in various other ways. Although forming a biological group they are of very diverse taxonomic affinities, indicating several independent evolutionary approaches to nematode parasitism. Genera include *Harposporium*, *Acrostalagmus* and *Meria* (hyphomycetes); *Nematoctonus* (basidoimycetes); *Catenaria*, *Myzocytium* and *Haptoglossa* (phycomycetes). Endozoic parasites have not been considered as promising fungi for use in attempts to control plant-pathogenic nematodes, initially at least because they were thought to be obligate parasites from which it would be impracticable to bulk up inoculum (Duddington, 1957). Many species have now been grown in pure culture (Barron, 1977). Giuma & Cooke (1974) investigated the possibility of reducing populations of nematodes in soil by adding conidia of *Nematoctonus* but the attempt was not successful.

FUNGAL PARASITES OF HETERODERA, GLOBODERA AND MELOIDOGYNE

Most of the attempts made with trapping fungi to control plant-parasitic nematodes were made against root-knot or cyst-forming nematodes of the genera *Meloidogyne* and *Heterodera*. The trapping fungi are adapted to capture vermiform nematodes which in these genera are the migrant larvae and the males. They are not adapted to capture the swollen females or the egg masses, which are the major phases of the life-cycle in these genera. There seems, in fact, only a single record of a classical nematophagous fungus parasitizing these phases: Korab's finding in 1929 of *Arthrobotrys oligospora* in cysts (egg masses protected within the dead female cuticle) of *Heterodera schachtii*.

Parasitism of eggs within the cyst of *H. schachtii*, or parasitism of cysts as it is often conveniently though imprecisely expressed, was studied in the Ukraine, Moravia and Germany over the period 1929–34, and substantial degrees of parasitism were reported. Individual species of fungi were accepted as the principal egg parasites, notably the fungi now known as *Phialophora malorum* and *Metarrhizium anisopliae*. Extensive studies recently made in Cambridge, Reading and Rothamsted on *H. schachtii* and *H. avenae* have failed to reveal either of these species. Absence of *M. anisopliae* is considered of especial interest because this is one of the principal insect-pathogenic fungi: there is now very little evidence that any insect-pathogenic fungus attacks nematodes or vice versa.

At Cambridge, cysts, principally of *H. schachtii* but also of *H. avenae* and *Globodera rostochiensis* were examined by microscopical observations of the egg content (Willcox & Tribe, 1974; Bursnall & Tribe, 1974; Tribe, 1977a, 1979). Each cyst examined was classed as full or empty, and the full cysts were classed as healthy, partially diseased (some eggs parasitized) or substantially diseased (most eggs parasitized). Representative parasitized eggs were then individually transferred to development media under close microscopical control and the fungus which grew out was identified and characterized. Development media include cornmeal agar, water drops covered with paraffin oil, or the cyst may be opened and the contents spread on a thin layer of water on a microscope slide and kept in a moist chamber. The last system is very liable to rapid over-growth by moulds. A procedure of this nature is regarded as essential for recovery of egg parasites because the simple direct plating out of surface-sterilized cysts, done in previous work, will permit the recovery of saprophytes growing on mucilage present between the eggs. Cysts, mostly of *H. schachtii*, were examined from many places and overall results are set out in Table 2. These results have been further analysed by Tribe (1977a, 1979). Most cysts were obtained from fields of sugar beet grown in rotation, but some of the German cysts were sent from mono-culture experimental plots where beet had been grown continuously for 11 years. Five egg pathogens were isolated from eggs of nearly half of the diseased cysts on samples from normally rotated beet and from 62 % of samples in the mono-culture. Miscellaneous fungi were isolated from some 10 % of normal samples but from few cysts in the mono-culture. Most other diseased cysts contained eggs which were 'lysed, shrivelled, coagulated or decayed', a category in which the eggs did not show evidence of a clear-cut occupation by fungal hyphae or uniform bacterial cells, but that of death of the egg content or lysis of a primary invading fungus usually followed by secondary organisms (decay). There was no evidence of primary bacterial parasitism of eggs. Sometimes eggs and larvae contained excessive amounts of oil; the cause of this disorder, which is of variable severity but can often completely destroy the egg content (Pl. 3A), is not known. It may be a physiological disorder or is perhaps virus induced.

The principal egg parasite (Pl. 3B, G, H) is *Verticillium chlamydosporium*, a well-known fungus widely distributed in soils. The second species is a sterile mycelium which grows very slowly in agar culture (less than 1 cm/month at laboratory temperatures), and termed the 'contortion fungus' because infected larvae are fixed in a contorted form (Pl. 3C). They are filled with mycelium and are very brittle. Three minor parasites are *Cylindrocarpon destructans*, a primary colonizer of senescent roots and sometimes pathogenic to roots of herbaceous and woody plants, a sterile fungus characterized by forming crystals in cornmeal agar and a fungus somewhat similar to, but distinct from, *Phialophora malorum* described by early workers. Two representative isolates of this fungus are known to

Table 2. *Overall analysis of extent and causation of disease in cysts and females of* Heterodera, *based on examination of 104 populations*

	H. schachtii			H. avenae	
	England and continental Europe	Germany (mono-culture)	USA	England	Germany (mono-culture)
No. of full cysts examined and (parentheses) % of diseased cysts	3164 (17·2)	580 (30·9)	384 (11·4)	177 (13·0)	115 (37·4)
Substantially diseased cysts (% diseased cysts)	52	68	39	61	81
Partially diseased cysts (% diseased cysts)	48	32	61	39	19
Cause or symptom of disease (% of diseased cysts)					
Recognized egg pathogens	47	62	0	61	77
Verticillium chlamydosporium	24	27	—	22	28
'Contortion fungus'	9	18	—	39	49
Cylindrocarpon destructans	5	8	—	0	0
'Crystal-forming fungus'	5	5	—	0	0
'Black yeast'	4	4	—	0	0
Miscellaneous fungi	10	2	0	0	5
Oiliness	5	<1	0	0	0
Lysed, shrivelled, coagulated and decayed	33	25	0	22	16
No. of females examined and (parentheses) % of diseased females	609 (9·0)	64 (0)	—	—	—
Cause or symptom of disease (% of diseased females)					
Specialized pathogens of females	91	0	—	—	—
Catenaria auxiliaris	67	—	—	—	—
'Wide-hyphal fungus'	24	—	—	—	—
Miscellaneous fungi	2	0	—	—	—
Oiliness	4	0	—	—	—

be *Exophiala pisciphila*, but it is not certain that all isolates are referable to *E. pisciphila*, and the fungus is denoted as a 'black yeast' (Pl. 3D).

In the few cysts of *Heterodera avenae* examined, *V. chlamydosporium* and the 'contortion fungus' were both found (Table 2). Graham & Stone (1975) found that nearly 40% of new cysts from Abingdon were diseased, mostly being occupied by fungi which were not identified. Chlamydospores of *V. chlamydosporium* were found amongst the eggs on some cysts. At Rothamsted, Kerry (1975) found that *V. chlamydosporium* was common in females and in new cysts of *H. avenae*, often killing all the eggs. He also found *C. destructans* in females, although this fungus killed few eggs.

The distinction between cyst and female in *H. schachtii* and *H. avenae* is not sharp. The female, on the host root, is white. Eggs are formed until the body is virtually filled, when the female dies because the internal organs degenerate and the cuticle becomes tanned, colouring first yellow and then brown. A practical distinction between female and cyst is made on the appearance of the yellow colour, but white nematodes may be found with eggs containing coiled larvae and yellow nematodes with eggs containing no developed larvae. It is not yet certain how soon females may be invaded by the fungi noted above. Apart from their activity as egg parasites, these fungi appear to be unspecialized saprophytes quite unrelated to one another.

Females of *H. schachtii* are principally parasitized by a fungus which has not yet been grown in artificial culture and is probably an obligate parasite. This was discovered by Kühn in 1877 as *Tarichium auxiliare* but largely because few females, contrasted with cysts, have been examined for parasites until recently, it remained almost unknown until it was re-studied following examination of the females of *H. schachtii* (Tribe, 1977b). Tribe transferred the species to the genus *Catenaria*, since it closely resembles *Catenaria spinosa*, a parasite of the eggs of the midge genus *Chironomus*. It is, however, quite possible that neither *C. auxiliaris* nor *C. spinosa* is really congeneric with the type species of *Catenaria, C. anguillulae*. *C. auxiliaris* invades the female to form a rhizomycelium which on maturation develops into strings of globose precursor sporangia. These mature either into zoosporangia or resting sporangia (Pl. 3E, F). A second, probably obligate, parasite, first referred to as '*Entomophthora*-like' and later as a 'wide-hyphal' fungus was found once only in females of *H. schachtii*. The smooth resting spores or sporangia failed to germinate and the life-cycle could not be worked out.

By contrast, females of *H. avenae* have been shown by Kerry (1974, 1975) and Kerry & Crump (1977, 1980) to be frequently parasitized by a fungus first known as '*Entomophthora*-like' and now as *Nematophthora gynophila*, the type species of a new genus. This produces zoospores from wide hyphae in a manner similar to *Leptolegniella*. These workers found *C. auxiliaris* infrequently on females of *H. avenae*, and a third fungus, a lagenidiaceous fungus, whose life-cycle is incompletely known, rarely in females (Kerry & Crump, 1980). Kerry & Crump (1977) have shown that *N. gynophila* is pathogenic to *H. schachtii, H. goettingiana, H. trifolii, H. carotae* and *H. cruciferae*, but not *Globodera* (*Heterodera*) *rostochiensis*. *C. auxiliaris* has not been recorded in *G. rostochiensis*, but has been found in laboratory populations of the closely related *G. pallida* (Kerry, Jenkinson & Crump cited by Jones (1976)).

There is general acceptance that *G. rostochiensis* has no effective natural enemies, at least in temperate countries (Jones, 1974), and evidence in support of this was provided by Willcox, who found no egg parasites in 5078 cysts examined from 20 populations (Willcox & Tribe, 1974). This was further confirmed by the experience of Kerry & Crump (1977). The diverse fungi reported from cysts of *G. rostochiensis* by van der Laan in 1956 were probably saprophytes, as van der Laan had plated whole cysts on to agar with no microscopical control (Tribe, 1977a). Latterly, one population of cysts of *G. rostochiensis* from near Cologne was found to contain many fungus-filled eggs (Goswami & Rumpenhorst, 1978). In this preliminary study, plating, but without microscopical control, resulted in isolation of *Fusarium oxysporum* and *F. solani*. Neither species could be transferred to eggs of healthy cysts on young potato plants, but inoculation of the cysts with fungus-filled eggs resulted in substantial parasitism. The fungus-filled eggs were accepted, though without experimental evidence, as being parasitized by one unknown fungus.

Studies of population dynamics of *G. rostochiensis* show that this nematode behaves in a manner predictable from simple mathematical models (Jones, 1974). Those of some (though not all) populations of *H. avenae* behave less predictably, as if influenced by an effective enemy (Jones, 1974; Graham & Stone, 1975). Perry (1978) proposed a model for the effects of a fungal parasite ($= Nematophthora$ *gynophila*) on densities of *H. avenae*. The model predicted population decline and attainment of low stable equilibrium densities and a good fit to certain field data was found. Assumptions for the model included production of some 4000 resting spores by the parasite in each infected female to provide the means of infection in subsequent years. These spores were assumed to be randomly distributed in the soil. It is noteworthy that although it must be regarded as certain that resting spores germinate under field conditions, neither those of *N. gynophila* nor of *C. auxiliaris* have yet been seen to germinate. The female nematode infected with *N. gynophila* or *C. auxiliaris* (at a sufficiently early age) rapidly disintegrates and disappears as a discrete unit. Kerry therefore recommends that observations of disease of females in roots must be made at frequent (weekly) intervals during the summer in order that the true extent of parasitism can be estimated.

Study of parasitism in *Meloidogyne* has mainly been restricted to nematode-trapping fungi. Certainly, in some hosts, *Meloidogyne* females are entirely buried in the tissues and are thus well protected against soil-borne parasites. On other hosts *Meloidogyne* behaves much as does *Heterodera*, although the egg masses are not enclosed within a cyst. In an early abstract, Linford & Oliveira (1938) listed 52 enemies of *Meloidogyne*, made up of 17 nematophagous fungi, 1 egg parasite (*Penicillium* sp.), 1 protozoon, 24 predacious nematodes, 6 mites and 3 predacious tardigrades. Linford and his colleagues, as already noted, restricted their studies to nematophagous fungi and only recently has work with other organisms commenced.

During a study of root-knot nematodes in certain Californian peach orchards, high populations of *Meloidogyne* which were expected on roots of peach trees were not found and suppression of the nematode by a parasite was suspected. This led to the discovery of a parasite of egg masses of *Meloidogyne* by Stirling & Mankau (1978). Although this fungus was described as a new species of *Dactylella* (*D. oviparasitica*) and thus placed in a genus which has often been used for classical nematode-trapping fungi, it differs in several respects from the latter, notably in never having been seen to form any kind of nematode trap. *D. oviparasitica* was isolated directly from *Meloidogyne* egg masses collected in the field or from egg masses on tomato seedlings grown in field soil. It parasitized most of the eggs in the relatively small masses (300–400 eggs) produced on peach roots but could manage rarely more than half the eggs in large egg masses (1000–1500 eggs) produced on tomato or grapevine roots (Stirling, McKenry & Mankau, 1979). Several species of nematode-trapping fungi were found in the peach orchard soils, but were not isolated from eggs, and Stirling, McKenry & Mankau estimated that their activity (on larvae) was unimportant in regulating *Meloidogyne* populations. Occasionally *D. oviparasitica* parasitized *Meloidogyne* females, particularly on hosts where the nematodes produced eggs relatively slowly. Stirling and his colleagues considered that *Meloidogyne* in the peach orchards was under natural biological control by *D. oviparasitica*. Although *D. oviparasitica* has not been recorded in *Heterodera* eggs, host range tests made by Stirling & Mankau (1979) showed that they would parasitize eggs of *H. schachtii* and of *Tylenchulus semipenetrans*, *Diplenteron* sp. and *Acrobeloides* sp. if the eggs of these nematodes were laid upon agar cultures of the fungus.

PARASITES OF EGGS OF OTHER NEMATODES

Barron (1977) stated there were very few fungi known to specialize as parasites of nematode eggs. He only discussed a single species, *Rhopalomyces elegans*, which invades eggs (of nematodes in general) on the surface of agar cultures. In soils, eggs of most nematodes are widely scattered and no-one has studied parasitism of eggs in the soil environment. Lysek, however, has studied parasitism of eggs of the animal parasite *Ascaris* by placing small piles of these eggs on to soil samples, so as to permit invasion from the soil. Lysek (1963) first noted that eggs were destroyed especially by *Fusarium redolens* and a *Cephalosporium* species, and later Lysek (1966) cited as principal parasites *F. redolens*, *F. javanicum* var. *radicicola*, *F. oxysporum* var. *orthoceras* and *Paecilomyces marquandii*, together with an unidentified bacterium. In another study (Lysek, 1968) he made a quantitative estimate of egg penetration by fungi. In piles of eggs laid on soil samples kept at 18–22 °C he found medium to slight penetration in 21·5 % of the 200 samples after 2 weeks; by 5 weeks 29 % of samples showed medium to slight penetration and 16 % showed strong penetration. Further species cited by Lysek (1978) to perforate egg shells and enter eggs were *Verticillium chlamydosporium*, *V. bulbillosum*, *Mortierella nana*, *Paecilomyces lilacinus*, *Acremonium bacillosporum* and *Helicoon farinosum*.

PROTOZOAN DISEASES

An amoeboid organism, described by Zwillenberg as *Theratromyxa weberi* is known to engulf and destroy the larvae of *Heterodera* in cultures. This was re-studied by Sayre (1973). As prey, Sayre used larvae of *Meloidogyne incognita*, *Heterodera trifolii*, and adults and larvae of *Aphelenchus avenae*, *Aphelenchoides rutgersi* and *Rotylenchus reniformis*. Sayre examined the parasitism of *T. weberi* on *M. incognita* in greenhouse tests and showed it was a poor biological control agent.

Canning (1973) reviewed several protozoan interactions with nematodes. Most were of minor importance but she regarded *Bodo caudatus* and *Duboscquia penetrans* as deserving of investigation as controlling agents. *B. caudatus* is able to aggregate in large numbers around the excretory pores, anus and mouth of nematodes. *D. penetrans*, after 35 years as a protozoon, is now generally accepted as a prokaryotic organism and is probably best placed as an actinomycete, as noted below.

BACTERIAL DISEASES OF NEMATODES

Two kinds of bacterial infection are now known. The first was originally described by Thorne (1940) as a microsporidian protozoon which he named *Duboscquia penetrans*. It was found in the vermiform root-lesion nematode *Pratylenchus penetrans*, occurring in 66 % of one population examined and in 28 % of another. Thorne considered the parasite was specific to *Pratylenchus*, noting that there were 21 other species of diverse nematodes associated with the *Pratylenchus* in the soils from which they came, but none was infected. Later, however, the parasite was found by various investigators in *Aphelenchoides*, *Dolichodorus*, *Rotylenchus*, *Tylenchorhynchus*, *Xiphinema* and larvae of *Meloidogyne* in several parts of the world (Canning, 1973). After studying this organism Mankau and his colleagues recognized its prokaryotic nature and transferred it to *Bacillus* as *B. penetrans* (Mankau, 1975; Mankau & Imbriani, 1975; Mankau & Prasad, 1977). Mankau himself recognized that the filamentous thalli found in the host were not typical of *Bacillus* species and Sayre & Wergin (1977) subsequently accepted that the affinities of the organism were probably with the Actinomycetales. They believe that it is an endospore-forming actinomycete whose exact systematic placing requires further research, and denoted it as 'BSPN' (bacterial spore parasite of nematodes). They drew attention to its close similarity with a parasite of Daphnidae, *Pasteuria ramosa*, whose life-cycle had been described by Metchnikoff in 1888 (Sayre, Wergin & Davis, 1977). The adhesive endospore of the actinomycete penetrates the cuticle of the nematode and develops in the pseudocoelom into a roughly spherical mycelial thallus, which breaks up and spreads throughout the host. Endospores are formed in cells on the periphery of the thalli and liberated into the soil upon death of the nematode (Fig. 3). Mankau evaluated this disease of nematodes as etiologically identical with the 'milky diseases' of beetle larvae caused by *Bacillus popilliae* and *B. lentimorbus*, species which are used in the USA for biological control of the Japanese beetle in grassland (De Bach, 1964).

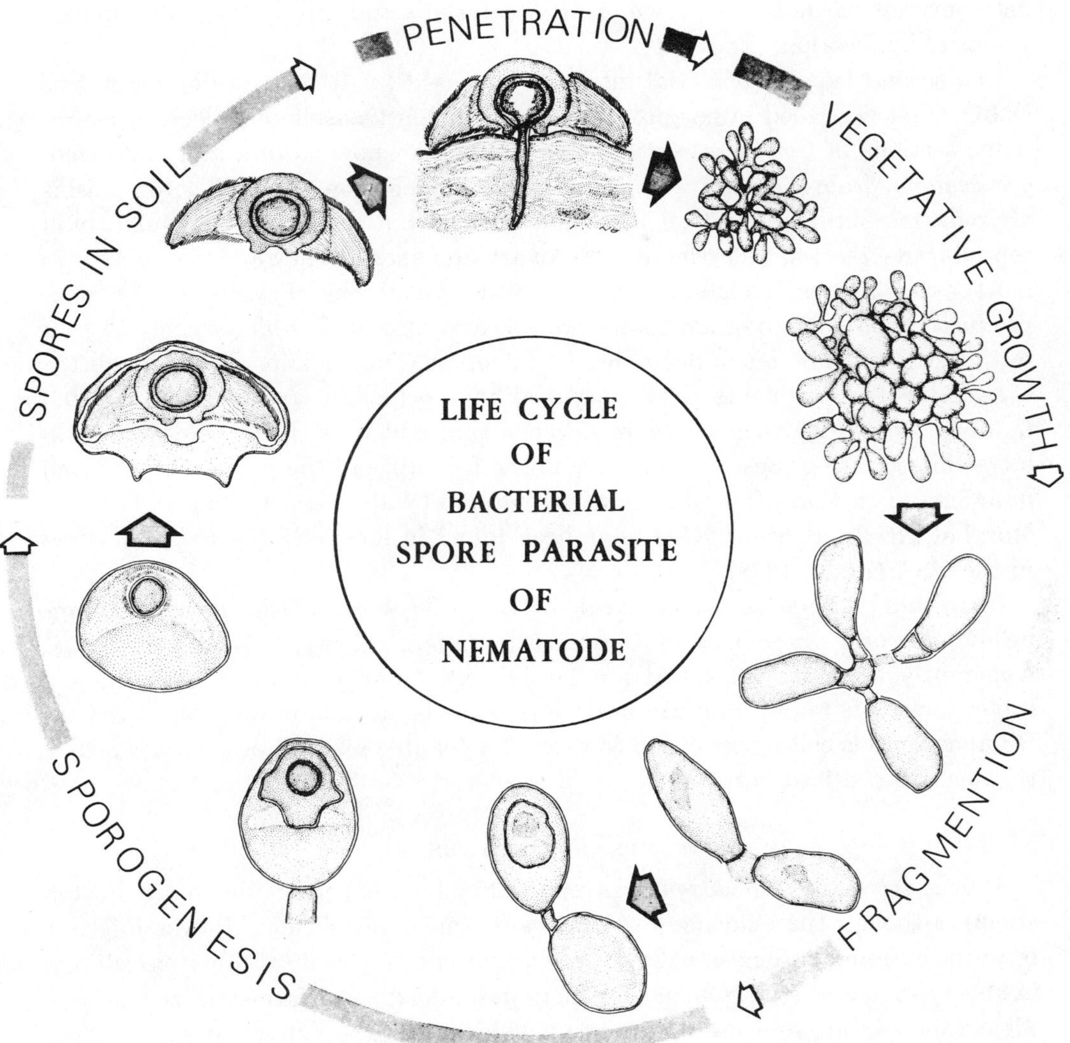

Fig. 3. Life-cycle of the bacterial spore parasite of nematodes. From R. M. Sayre & W. P. Wergin (1977), by kind permission of The American Society for Microbiology.

Larvae of *Meloidogyne* become infected as they pass through soil containing the endospores. Females dissected from galls in host plant roots usually produce no eggs and contain about 2×10^6 spores (Mankau, 1975). Presumably these females develop from slightly infected larvae which enter the roots. Mankau & Prasad (1977) examined the host range of a population of the actinomycete from *M. javanica* and showed that *M. arenaria, M. hapla, M. incognita* and *Pratylenchus scribneri* were good hosts, but neither larvae nor adults of the following species were infected: *Aphelenchoides* sp., *Aphelenchus avenae, Aporcelaimus* sp., *Ditylenchus dipsaci, Heterodera schachtii, Trichodorus christii, Tylenchorhynchus claytonii* or *Xiphinema index*. They noted its host range was more limited than would be apparent from the literature, which raises the question of strain specificity. The actinomycete has not been grown on artificial media and this undoubtedly hinders practical application.

The second type of bacterial infection is caused by Rickettsia-like organisms (RLO), first described in nematodes from ultra-thin transmission electron microscope sections of *Globodera rostochiensis* (a Bolivian population), and *Heterodera goettingiana*, from England (Shepherd, Clark & Kempton, 1973). The organisms are rods, measuring up to $3 \times 0\cdot5$ μm, and are distributed especially abundantly in reproductive tissues, occurring in both ovary and sperm. They are also present in most other tissues, including intestine and hypodermis. Fertilized eggs were produced even when ovarian tissues were heavily infected, which suggested that fertility may not be much decreased by infection. Transmission was through the eggs. Recent experiments have suggested that penicillin treatment of infected *G. rostochiensis*, which resulted in degenerating and dead RLO being found in ultra-thin sections, caused an increase in the fecundity of the population but had no influence on that of an uninfected population (Walsh, Lee & Shepherd, 1979). Morphologically identical RLO have been found in larvae of *Heterodera glycines* in the USA (Endo, 1979).

Ultra-thin sections have also revealed RLO in *Xiphinema index* recovered from yellows-diseased grapevines in Germany (Rumbos, Sikora & Nienhaus, 1977). Apparently the same RLO had previously been found in ultra-thin sections of leaves and roots taken from diseased vines, and the nematode was considered by Rumbos and his colleagues in the context of a possible plant disease vector rather than as a parasitized host.

VIRUS DISEASES

A virus disease of *Meloidogyne* was reported by Loewenberg, Sullivan & Schuster (1959), although the evidence for viral causation is insufficient. It was inferred from the extremely sluggish behaviour of larvae entering roots of tomato seedlings in an experimental study. Sluggish nematodes, added to surface-sterilized eggs of *Meloidogyne* in an aqueous environment, transmitted the behaviour to the eggs, the larvae hatching from the eggs being slow and jerky in movement and many died. Examination of larvae under the light microscope failed to reveal microorganisms, but the larvae appeared to be highly vacuolated or filled with unusually prominent oil-like globules. Unfortunately no further work on this interesting system appeared.

The swarming phenomenon in nematodes, known in many species and demonstrated by congregation of nematodes in masses in aqueous suspension, has recently been claimed to be viral in origin (Ibrahim, Joshi, & Hollis, 1978). The phenomenon results from stickiness of the cuticle. Electron micrographs of *Tylenchorhynchus martinii* showed high numbers of symmetrical virus-like inclusion bodies in the nematode tissues. None was present in healthy, non-swarming *T. martinii*. Ibrahim *et al.* (1978) accepted that swarming was a disease of nematodes associated with the virus-like bodies, which were similar to the viruses causing cytoplasmic polyhedrosis in insects.

NEMATODE PATHOLOGY AS A DISCIPLINE

The present is a period of great activity in the field of nematode pathology. As recently as 1977, Barron could give 'The Nematode-destroying Fungi' as the title of his book on the fungi parasitic on vermiform nematodes. From about that time work with the important genera of cyst- and root-knot nematodes, *Heterodera* and *Meloidogyne*, has shown that there are two further widespread and abundant groups of nematode-destroying fungi: the apparently non-specialized and unrelated parasites of eggs in cysts or egg-masses, and the rather specialized probably obligate zoosporic fungi which destroy the females. The nature of '*Duboscquia penetrans*' has been elucidated and rickettsia-like organisms discovered in nematodes. There have been indications of virus diseases, and since viruses must be expected in nematodes as well as in other groups of organisms, more cases are likely to follow. The time is now ripe for the searching out and investigation of symptoms of diseases and disorder in nematodes. The oiliness symptom (if it be a single disorder) is a case in point. It is noteworthy that Loewenberg *et al.* (1959) in their preliminary studies indicating a virus causing sluggish behaviour in *Meloidogyne* larvae wrote of larvae filled with oil-like globules. Although not normally a frequent symptom, according to present information, occasional populations of *Heterodera* examined by the author have consisted of nearly all oily eggs.

Very little detailed investigation has been made of the state of health and disease in natural and field populations of nematodes. This is partly because of the time-consuming nature of such studies, but it is important to know what this state is in normal populations, so that any given situation which appears of especial interest may be compared with 'norms'. Certainly it is of especial interest to investigate those populations of plant-parasitic nematodes where expected severity in plant disease fails to materialize.

Of the various micro-organisms now known to be active against plant-parasitic nematodes, only *V. chlamydosporium* and perhaps *D. oviparasitica* at present seem likely candidates for direct addition. An early attempt with *V. chlamydosporium* (Willcox & Tribe, 1974) was not successful: here the fungus was added as conidial suspensions. These will spread within soil pores more readily than the chlamydospores, but the latter with their far greater size would appear better sources of infection. However, with reference to *H. schachtii*, beet mono-culture in the presence of *V. chlamydosporium* for 11 years encouraged *V. chlamydosporium* only sufficiently to account for disease in some 8 % of all full cysts examined and only about two-thirds of these were 'substantially diseased'. The specificity of *V. chlamydosporium* has not yet been fully evaluated; it has been found in eggs of *H. avenae* as well as *H. schachtii*, but has not yet been recovered from *G. rostochiensis* or *Meloidogyne* spp. However, it is strongly parasitic to *Ascaris* eggs (Lysek, 1978) and it has been recorded on eggs of a snail. These pointers suggest that it could be a generalized egg parasite and may be particularly prominent in eggs of *Heterodera* because these are presented in large packets. Whether *V. chlamydosporium* is distributed reasonably evenly throughout agricultural soils is not certain; although commonly found it was notably absent from samples studied by the author from USA (California, Colorado & Michigan). The 'contortion fungus' is known from *H. schachtii*, *H. avenae* and from some of the few cysts of *H. glycines* examined by the author (from Illinois, USA). So far, however, this fungus is incompletely known, no spores having ever been found. Although *D. oviparasitica* has not yet been found in populations of *Heterodera*, pathogenicity tests conducted by Stirling & Mankau (1979) showed that it parasitized eggs of *H. schachtii* when the eggs were placed upon cultures of the fungus. How far such tests relate to conditions in the field is uncertain. Placing an egg on an agar surface occupied by a fungus will give the fungus a much better opportunity for parasitism than if the egg is surrounded by a non-nutrient medium such as agricultural soil. Fungus hyphae will be actively growing and, if potentially capable of invading eggs, will likely do so, whereas in agricultural soil the fungus must be presumed present as spores which must be stimulated to germinate and invade the egg with vastly less inoculum potential. The part played by mucilage in cysts may, however, be analogous to that played by agar. If a fungus grows vigorously in the mucilage of a young cyst it will obviously have every opportunity of invading eggs (Pl. 3 H). If the mucilage has disappeared or is pre-colonized by a fungus unable to parasitize eggs, the opportunity is likely lessened.

Eggs of *Heterodera* are also occupied, though less frequently, by fungi evaluated as 'minor pathogens' or 'miscellaneous'. *Cylindrocarpon destructans* is a minor pathogen; it is known from eggs of *H. schachtii* and *H. avenae* but its primary activity in the soil environment is colonization of senescing plant roots. *Fusarium* species have been classed by the author as 'miscellaneous' as they have only occasionally been found in eggs of *H. schachtii* and *H. avenae*. They were found frequently in the piles of *Ascaris* eggs by Lysek. The taxa recorded by Lysek are strains of *Fusarium oxysporum* and *F. solani* as classified by Snyder and Hansen, and both are closely related to *C. destructans*. They are amongst the most abundant of all soil-borne fungi, numbering thousands of propagules/g in agricultural soils. From their activities on plant roots, egg parasitism must be a minor occupation. Parasitism of eggs by these fungi, or by species of *Penicillium*, *Paecilomyces* or *Mortierella*, all amongst the most abundant of soil fungi, is very likely dependent on some deficient condition in the egg or contributory factor of the environment in which it is placed. Whether strains of these fungi 'strongly pathogenic to eggs' can arise is very questionable. An egg is a small structure, and may contain cytoplasm closely appressed to the walls or a larva separate from the walls. As a single unit it contains hardly sufficient nutrient to support a filamentous fungus to the stage of reproduction, although certain lower fungi can be so supported. When the eggs are in cysts, masses or piles a filamentous fungus emerges from the first parasitized egg to penetrate the adjacent egg, and so on through the rest of the eggs. It is noteworthy that in both *Heterodera* and *Meloidogyne* egg-masses, parasitic fungi often fail to complete their task. About half of all *Heterodera* cysts examined show only partial destruction (less than 50 % of eggs destroyed). This figure applies equally to populations overwintered in the field, although periods of months should represent more than enough time for even leisurely progress through the eggs of a cyst. The role of mucilage in the cyst may be important here, and early colonization of this may greatly further, or even be necessary for, complete destruction of the egg content.

By contrast to the apparently rather unspecialized nature of egg parasites, parasites of females seem much more specific. *Catenaria auxiliaris*, despite its highly characteristic resting spores which make recognition an easy matter, has only been reported from cyst-nematodes (with the single exception of its identification in a larva of the beetle *Scolytus scolytus* (Doberski & Tribe, 1978)). *Nematophthora gynophila*, also a highly characteristic fungus, is a new species so far known only from cyst-nematodes. Present evidence suggests that *C. auxiliaris* is common in *H. schachtii* and rare in *H. avenae*, and *N. gynophila* rare in *H. schachtii* and common in *H. avenae*, but despite a modicum of study on these species in the last few years, much more information on distribution is necessary before this suggestion is satisfactorily confirmed. *N. gynophila* parasitizes females of all other *Heterodera* species tested by Kerry & Crump (1977) (happily, they were tested in the soil environment and not on agar plates) but it did not parasitize *Globodera*. Such tests have not been made for *C. auxiliaris*.

If it is proven that, as Kerry and colleagues believe, *N. gynophila* does limit the multiplication of *H. avenae* in (parts of) England and exerts a natural biological control, then the case for spreading it, probably in natural soil, to regions where it is absent will be a good one. The evidence of *C. auxiliaris* reducing *H. schachtii* populations in sugar beet is less strong. Both fungi are zoosporic and therefore need water for rapid spread; there is evidence that their activities are inhibited in dry soils.

The rarity of parasitism in *G. rostochiensis* is very interesting. This nematode was accepted in *Heterodera* for many years, but the sub-genus *Globodera* was raised to generic rank in 1976 and *rostochiensis* made the type species. The chief difference from *Heterodera* is in the cyst shape, which is nearly globose and results from absence of a vulval cone. No eggs are extruded into the soil by *G. rostochiensis* and the vulval aperture in the cyst is small. Further, there is a layer of cuticle in the female *G. rostochiensis* additional to those present in *Heterodera* (Shepherd, Clark & Dart, 1972).

With reference first to the less specific type of parasitism in eggs, it is possible that both morphological features in *Globodera* of reduced apertures in the cyst and the extra layer of cuticle restrict parasitism by fungi. *V. chlamydosporium* spores have been shown to penetrate the cyst wall in *H. avenae* and attack the eggs within (Kerry, Crump, Mullen & Clark, cited by Jones, 1978), but no cyst-penetration study has been made with *G. rostochiensis*. Kerry, Jenkinson & Crump (cited by Jones, 1976) have noted that eggs of *G. rostochiensis* are susceptible to *V. chlamydosporium*. We do not know why *G. rostochiensis* is resistant to egg pathogens.

Concerning the more specific pathogens of females, it seems likely here that strain differences in these pathogens may influence host range. The present picture of *C. auxiliaris* being a principal pathogen of *H. schachtii* and *N. gynophila* of *H. avenae* may be a strain rather than a species character. As already noted, all species of *Heterodera* tested by Kerry & Crump were susceptible to *N. gynophila*, and *C. auxiliaris* has been found in *H. avenae* and also in *H. carotae* (Cayrol, personal communication). Both species, therefore, are parasites of *Heterodera*, but strain differences may determine their virulence in any given species or strain of *Heterodera*. It is generally accepted that *G. rostochiensis* originated from S. America, and it would certainly seem that a search for *C. auxiliaris* and *N. gynophila* strains on females of *G. rostochiensis* in its centre of origin is an immediate next step.

CONCLUSION

Many more pathogens are now known on plant-parasitic nematodes than even a few years ago. Others may be expected, including viruses and perhaps physiological and genetic causation of disease. Our knowledge of the distribution and host range of nematode diseases in nature is minimal. There is need for systematic study of cross-relationships between nematode pathogen, geographical area, immediate location and nematode host; in any given population of plant-parasitic nematodes the presence and frequency or absence of each pathogen needs to be established. Answers are needed to such questions as (1) which pathogens does *H. schachtii* carry in (the many regions of) the USA, (2) how widespread in host and geography is the BSPN actinomycete '*Dubosquia penetrans*', (3) which are the pathogens of *Globodera* in S. America and (4) how uniformly spread are pathogens of female *H. avenae* in English cereal fields? The science of nematode pathology is coming of age and a systematic use of it will guide us into practicable possibilities for biological control.

It is a pleasure to thank Dr Audrey Shepherd and Dr B. R. Kerry of the Nematology Department at Rothamsted Experimental Station and Dr J-C. Cayrol of I.N.R.A., Antibes, France for information on recent literature and for enjoyable conversations.

REFERENCES

BARRON, G. L. (1977). *The Nematode-destroying Fungi*. Guelph, Ontario: Canadian Biological Publications Ltd.

BURSNALL, L. A. & TRIBE, H. T. (1974). Fungal parasitism in cysts of *Heterodera*. II. Egg parasites of *H. schachtii*. *Transactions of the British Mycological Society* **62**, 595–601.

CANNING, E. U. (1973). Protozoal parasites as agents for biological control of plant-parasitic nematodes. *Nematologica* **19**, 342–8.

CAYROL, J-C. & FRANKOWSKI, J-P. (1979). Une méthode de lutte biologique contre les nématodes à galles des racines appartenant au genre *Meloidogyne*. *Pépiniéristes Horticulteurs Maraîchers – Revue Horticole* **193**, 15–23.

CAYROL, J-C., FRANKOWSKI, J-P., LANIECE, A., D'HARDEMARE, G. & TALON, J-P. (1978). Contre les nématodes en champignonnière. *Pépiniéristes Horticulteurs Maraîchers – Revue Horticole* **184**, 23–30.

COOKE, R. C. (1968). Relationships between nematode-destroying fungi and soil-borne phytonematodes. *Phytopathology* **58**, 909–13.

COOKE, R. C. (1977). *The Biology of Symbiotic Fungi*. Chichester: Wiley.

COOKE, R. C. & GODFREY, B. E. S. (1964). A key to the nematode-destroying fungi. *Transactions of the British Mycological Society* **47**, 61–74.

DE BACH, P. (1964). *Biological Control of Insect Pests and Weeds*. London: Chapman and Hall.

DOBERSKI, J. W. & TRIBE, H. T. (1978). *Catenaria auxiliaris* identified in a larva of *Scolytus scolytus*. *Journal of Invertebrate Pathology* **32**, 392–3.

DRECHSLER, C. (1937). Some hyphomycetes that prey on free-living terricolous nematodes. *Mycologia* **29**, 447–552.

DRECHSLER, C. (1941). Predaceous fungi. *Biological Reviews* **16**, 265–90.

DUDDINGTON, C. L. (1956). The predacious fungi: Zoopagales and Moniliales. *Biological Reviews* **31**, 152–93.

DUDDINGTON, C. L. (1957). *The Friendly Fungi*. London: Faber and Faber.

DUDDINGTON, C. L. (1962). Predacious fungi and the control of eelworms. In *Viewpoints in Biology*, vol. 1 (ed. C. L. Duddington and J. D. Carthy). London: Butterworths.

DUDDINGTON, C. L. & WYBORN, C. H. E. (1972). Recent research on the nematophagous hyphomycetes. *Botanical Review* **38**, 545–65.

ENDO, B. Y. (1979). The ultrastructure and distribution of an intracellular bacterium-like microorganism in tissue of larvae of the soybean cyst nematode *Heterodera glycines*. *Journal of Ultrastructure Research* **67**, 1–14.

ESTEY, R. H. & OLTHOF, T. H. A. (1965). The occurrence of nematophagous fungi in Quebec. *Phytoprotection* **46**, 14–17.

FOWLER, M. (1970). New Zealand predacious fungi. *New Zealand Journal of Botany* **8**, 283–302.

GIUMA, A. Y. & COOKE, R. C. (1974). Potential of *Nematoctonus* conidia for biological control of soil-borne phytonematodes. *Soil Biology Biochemistry* **6**, 217–20.

GOSWAMI, B. K. & RUMPENHORST, H. J. (1978). Association of an unknown fungus with potato cyst nematodes, *Globodera rostochiensis* and *G. pallida*. *Nematologica* **24**, 251–6.

GRAHAM, C. W. & STONE, L. E. W. (1975). Field experiments on the cereal cyst-nematode (*Heterodera avenae*) in south-east England 1967–72. *Annals of Applied Biology* **49**, 515–23.

IBRAHIM, I. K. A., JOSHI, M. M. & HOLLIS, J. P. (1978). Swarming disease of nematodes: host range and evidence for a cytoplasmic polyhedral virus in *Tylenchorhynchus martini*. *Proceedings of the Helminthological Society of Washington* **45**, 233–8.

JONES, F. G. W. (1974). Control of nematode pests, background and outlook for biological control. In *Biology in Pest and Disease Control* (ed. D. Price-Jones and M. E. Solomon). Oxford: Blackwell.

JONES, F. G. W. (1976). Nematology Department. *Rothamsted Report for 1975*, Part 1, 191–212.

JONES, F. G. W. (1978). Nematology Department. *Rothamsted Report for 1977*, Part 1, 171–91.

KERRY, B. R. (1974). A fungus associated with young females of the cereal cyst-nematode, *Heterodera avenae*. *Nematologica* **20**, 259–60.

KERRY, B. R. (1975). Fungi and the decrease of cereal cyst-nematode populations in cereal monoculture. *European Plant Protection Organization Bulletin* **5**, 353–61.

KERRY, B. R. & CRUMP, D. H. (1977). Observations on fungal parasites of females and eggs of the cereal cyst-nematode, *Heterodera avenae*, and other cyst-nematodes. *Nematologica* **23**, 193–201.

KERRY, B. R. & CRUMP, D. H. (1980). Two fungi parasitic on females of cyst-nematodes (*Heterodera* spp.). *Transactions of the British Mycological Society* **74**, 119–25.

LINFORD, M. B. & OLIVEIRA, J. M. (1938). Potential agents of biological control of plant-parasitic nematodes. *Phytopathology* **28**, 14.

LOEWENBERG, J. R., SULLIVAN, T. & SCHUSTER, M. L. (1959). A virus disease of *Meloidogyne incognita incognita*, the southern root-knot nematode. *Nature, London* **184**, 1896.

LYSEK, H. (1963). Effect of certain soil organisms on the eggs of parasitic roundworms. *Nature, London* **199**, 925.

LYSEK, H. (1966). Study of biology of geohelminths. II. The importance of some soil microorganisms for the viability of geohelminth eggs in the soil. *Acta Universitatis Palackianae Olomucensis* **40**, 83–90.

LYSEK, H. (1968). Biological liquidation of ascarid eggs in spring pasture soil. *Acta Parasitica Polonica* **15**, 263–7.

LYSEK, H. (1978). A scanning electron microscope study of the effect of an ovicidal fungus on the eggs of *Ascaris lumbricoides*. *Parasitology* **77**, 139–41.

MANKAU, R. (1975). *Bacillus penetrans* n. comb. causing a virulent disease of plant-parasitic nematodes. *Journal of Invertebrate Pathology* **26**, 333–9.

MANKAU, R. & IMBRIANI, J. L. (1975). The life cycle of an endoparasite in some tylenchid nematodes. *Nematologica* **21**, 89–94.

MANKAU, R. & PRASAD, N. (1977). Infectivity of *Bacillus penetrans* in plant-parasitic nematodes. *Journal of Nematology* **9**, 40–5.

NORTON, D. C. (1963). Iowa fungi parasitic on nematodes. *Proceedings of the Iowa Academy of Sciences* **69**, 108–17.

PERRY, J. N. (1978). A population model for the effect of parasitic fungi on numbers of the cereal cyst-nematode, *Heterodera avenae*. *Journal of Applied Ecology* **15**, 781–7.

RUMBOS, I., SIKORA, R. & NIENHAUS, F. (1977). Rickettsia-like organisms in *Xiphinema index* Thorne and Allen found associated with yellows disease of grapevines. *Zeitschrift für Pflanzenkrankheiten und Pflanzenschutz* **84**, 240–3.

SAYRE, R. M. (1973). *Theratromyxa weberi*, an amoeba predatory on plant-parasitic nematodes. *Journal of Nematology* **5**, 258–64.

SAYRE, R. M. & WERGIN, W. P. (1977). Bacterial parasite of a plant nematode: morphology and ultrastructure. *Journal of Bacteriology* **129**, 1091–101.

SAYRE, R. M., WERGIN, W. P. & DAVIS, R. E. (1977). Occurrence in *Monia rectirostris* (Cladocera: Daphnidae) of a parasite morphologically similar to *Pasteuria ramosa* (Metchnikoff, 1888). *Canadian Journal of Microbiology* **23**, 1573–9.

SCHENCK, S., KENDRICK, W. B. & PRAMER, D. (1977). A new nematode-trapping hyphomycete and a re-evaluation of *Dactylaria* and *Arthrobotrys*. *Canadian Journal of Botany* **55**, 977–85.

SHEPHERD, A. M., CLARK, S. A. & DART, P. J. (1972). Cuticle structure in the genus *Heterodera*. *Nematologica* **18**, 1–17.

SHEPHERD, A. M., CLARK, S. A. & KEMPTON, A. (1973). An intracellular micro-organism associated with tissues of *Heterodera* spp. *Nematologica* **19**, 31–4.

SOPRUNOV, F. F. (1966). *Predacious Hyphomycetes and their Application in the Control of Pathogenic Nematodes*. Israel Programme for Scientific Translations, Jerusalem. Available from the U.S. Department of Commerce, Springfield, Virginia 22151.

STIRLING, G. R. & MANKAU, R. (1978). *Dactylella oviparasitica*, a new fungal parasite of *Meloidogyne* eggs. *Mycologia* **70**, 774–83.

STIRLING, G. R. & MANKAU, R. (1979). Mode of parasitism of *Meloidogyne* and other nematode eggs by *Dactylella oviparasitica*. *Journal of Nematology* **11**, 282–8.

STIRLING, G. R., McKENRY, M. V. & MANKAU, R. (1979). Biological control of root-knot nematodes (*Meloidogyne* spp.) on peach. *Phytopathology* **69**, 806–9.

THORNE, G. (1940). *Duboscqia penetrans*, n.sp. (Sporozoa, Microsporidia, Nosematidae), a parasite of the nematode *Pratylenchus pratensis* (de Man) Filipjev. *Proceedings of the Helminthological Society of Washington* **7**, 51–3.

TRIBE, H. T. (1967). Practical studies on biological decomposition in soil: a simple technique for observation of soil organisms colonizing buried cellulose film. *The School Science Review* **167**, 95–112.

TRIBE, H. T. (1977*a*). Pathology of cyst-nematodes. *Biological Reviews* **52**, 477–507.

TRIBE, H. T. (1977*b*). A parasite of white cysts of *Heterodera*: *Catenaria auxiliaris*. *Transactions of the British Mycological Society* **69**, 367–76.

TRIBE, H. T. (1979). Extent of disease in populations of *Heterodera*, with especial reference to *H. schachtii*. *Annals of Applied Biology* **92**, 61–72.

WALSH, J. A., LEE, D. L. & SHEPHERD, A. M. (1979). Rickettsial parasites of cyst-nematodes. In *IX International Congress of Plant Protection, Washington, DC*. Abstract no. 914.

WILLCOX, J. & TRIBE, H. T. (1974). Fungal parasitism in cysts of *Heterodera*. I. Preliminary investigations. *Transactions of the British Mycological Society* **62**, 585–94.

EXPLANATION OF PLATES

PLATE 1

Habit of a constricting ring fungus in the soil environment. The Plate is reduced from a tracing made of a composite photo-micrograph at × 80. The fungus is growing over the surface of dead and living bacterial cells which have replaced a piece of buried cellulose film supported on a slide (see text). It has set many ring traps and captured 13 nematodes (arrowed) but many other nematodes are healthy and reproducing. Although established over the lower part of the residue, the fungus has not spread to the upper part.

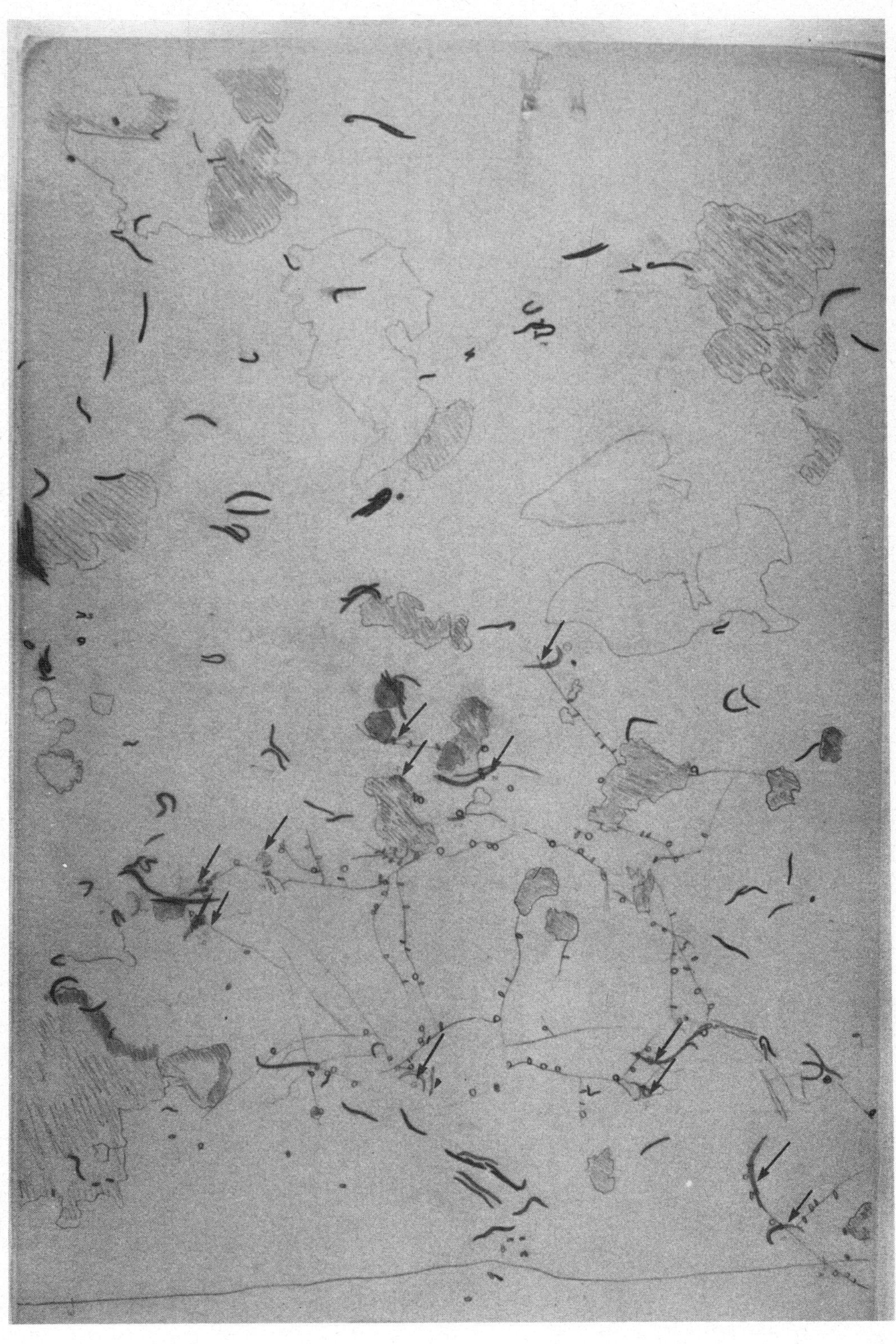

H. T. TRIBE

(*Facing p.* 100)

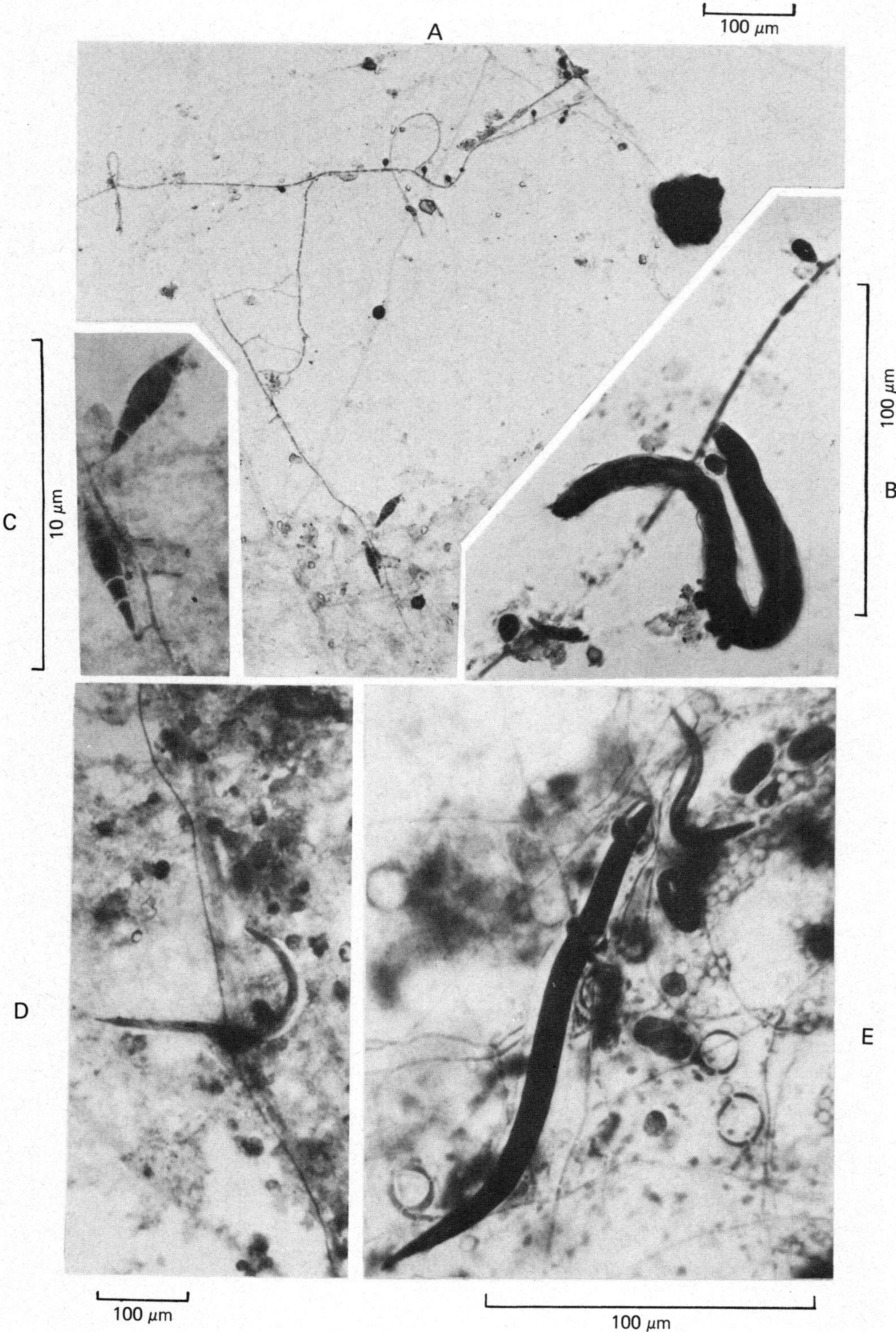

H. T. TRIBE

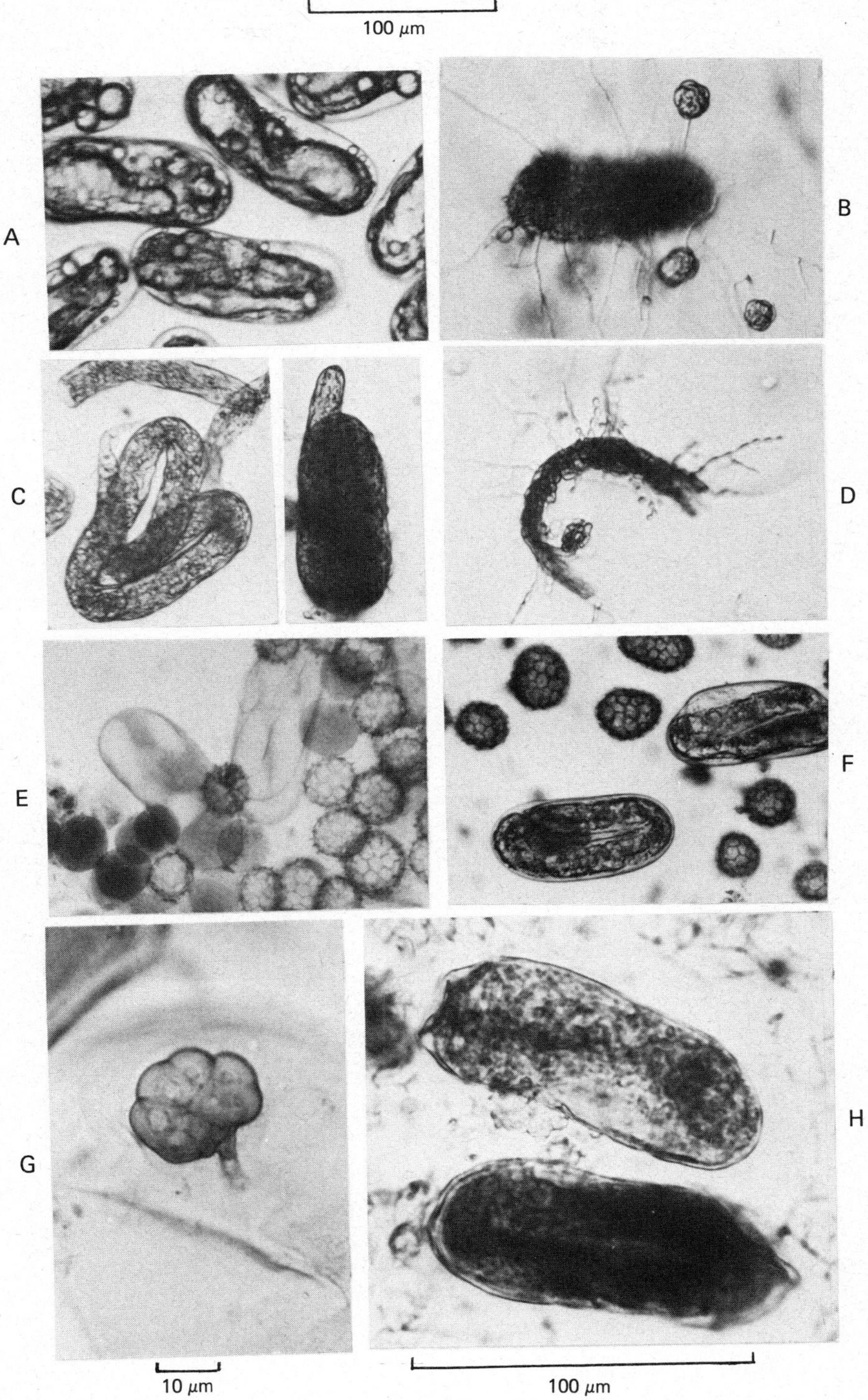

H. T. TRIBE

PLATE 2

Nematode-trapping fungi acting in the soil environment. Growth is over bacterial mush which has resulted from prolonged decomposition of cellulose film buried in soil on supporting glass slides (see text).

A. Adhesive knob-bearing fungus. This is extending from the bacterial residue (below) over the glass support where it bears a row of knobs. Two conidia are borne from a hyphal strand at the margin of the bacterial residue. From cellulose film buried in forest soil for 27 days.

B. Nematode held in two places by one adhesive knob. From another part of the same slide.

C. The two conidia enlarged. They indicate that the fungus is *Monacrosporium ellipsoideum*. They are borne on very short conidiophores which are in strong contrast to the long conidiophores typical on agar media.

D. Adhesive mycelial fungus growing as a single strand over bacterial residue. A nematode is trapped; the dark patch just below it represents the abundant mucilage secreted by this type of fungus. From cellulose film buried in clay soil for 80 days.

E. Constricting ring-bearing fungus over bacterial residue. One nematode is trapped in two places. Three unsprung constricting rings are clearly visible together with 1 small nematode and 3 eggs. From cellulose film buried in clay soil for 80 days.

PLATE 3

Cyst-nematode disease and pathogens.

A. Eggs destroyed by oily degeneration.

B. Egg filled with mycelium of *Verticillium chlamydosporium*. This has been placed for a few days in a thin layer of water in a humid chamber. The characteristic chlamydospores (see Pl. 3 G below) are borne on the emerging mycelium.

C. Hatched larva (left) and partially hatched larva (right) destroyed by the 'contortion fungus'. Preparation made from freshly opened cyst of *Heterodera*.

D. Hatched larva filled with mycelium of a 'black yeast'. It has been placed in a thin layer of water for 4 days in a humid chamber. Hyphae, some of which are moniliform, are growing out.

E. Female *Heterodera* destroyed by *Catenaria auxiliaris*. The fungus is represented by immature resting sporangia, without reticulations, and mature resting sporangia, each containing a single reticulate resting spore. The eggs are unparasitized.

F. Result of a late attack on a female nematode by *C. auxiliaris*. Some of the tissue of the old female is converted into resting sporangia but the pre-formed eggs are unparasitized and the cuticle has undergone tanning to give a normal cyst.

G. Single chlamydospore of *V. chlamydosporium* borne on a pedicel from mycelium which has lysed. The chlamydospore is multicellular and very thick-walled.

H. Eggs from a young cyst adhering to a root showing vigorous growth of *V. chlamydosporium* in the eggs and in the surrounding mucilage. A larva is visible in the lower egg, filled with mycelium